The Nobel for Accidents

Sandeep Chavan

Published by SANDEEP CHAVAN, 2025.

THE NOBEL FOR ACCIDENTS

First edition. October 10, 2025.

ISBN: 979-8232730949

Written by Sandeep Chavan.

Table of Contents

To the 2025 Nobel Laureates —

John Martinis, John Clarke, and Michel Devoret
—whose extraordinary experiment in silence revealed more than motion ever could.You reminded the world that when noise ends, nature speaks —and sometimes, coherence itself deserves the prize.

"The universe never rewards confusion — it rewards alignment.Every so-called accident is just nature remembering herself."

— Sandeep Chavan

Disclaimer

This book represents an **independent philosophical and scientific reconstruction** of reality through the framework of **Universal Energy Dynamics (UED)**.

It is not affiliated with, endorsed by, or representative of any established academic institution, research organization, or Nobel Committee. The names of scientists, awards, and historical events mentioned herein are used **purely for critical and educational discussion**.

The concepts, interpretations, and theories presented — including Structured Vacuum Energy (SVE), Ripple Coherence, and the Law of Consequence (LOC) — reflect the author's **original research, synthesis, and perspective**, derived from cross-disciplinary analysis and philosophical reasoning. They are intended to provoke reflection and inquiry, not to replace formal scientific education or experimental validation.

All experiments, historical accounts, and contemporary research references have been reinterpreted **within the context of the UED framework** to illustrate broader principles of coherence, alignment, and continuity in nature. Readers are encouraged to approach these interpretations with **open-minded skepticism** — questioning, testing, and extending them through their own observation and understanding.

The book's purpose is **intellectual exploration**, not prediction or prescription. It does not offer technical, medical, or psychological advice. Any resemblance to ongoing scientific work or institutional research is **coincidental in method but intentional in spirit** — to invite dialogue toward a more unified and coherent worldview.

"This book is not against science — it is science remembering itself."

— **Sandeep Chavan**

Author & Researcher

Opening Note

When I first read the list of Nobel Prizes in physics, I noticed something no one talks about — how many of them were, quite honestly, *accidents*. Not miracles, not prophecies — accidents. From X-rays discovered by a man who wasn't even looking for them, to quantum mechanics born out of panic over a blackbody curve, to superconductivity stumbled upon by someone who simply refused to warm the wires — the pattern was too consistent to ignore.

The universe, it seems, keeps rewarding human confusion — with medals.

But what if these so-called accidents weren't coincidences at all? What if nature, in her calm intelligence, was merely revealing coherence every time our chaos quieted down enough to notice?

That is the spirit of this book.

We live in a world that worships discovery, but rarely questions *how* discovery happens. Every time we hit a wall in theory, we patch it with a constant, an equation, or a clever interpretation — and when the patch holds, we celebrate. But the 2025 Nobel Prize — for observing macroscopic quantum tunneling and energy quantization in electric circuits — is the latest reminder that nature doesn't need our patches; she needs our patience.

Those superconducting circuits didn't "break" the barrier between classical and quantum physics. They simply aligned so well that the barrier itself dissolved. The scientists called it *quantum tunneling.* Under Universal Energy Dynamics (UED), it's just the field remembering its own coherence — the natural stillness beneath all motion.

I've written this book to show, with humor and humility, that many of humanity's proudest "discoveries" were really moments of accidental alignment. We kept naming the ripples but never looked at the water. We mistook the echo for the voice.

UED doesn't ridicule science — it restores its continuity. It tells us that nothing ever jumps, tunnels, collapses, or pops out of nowhere. It all unfolds through ripple coherence — steady, continuous, and patient — waiting for the observer to stop disturbing it.

This book is a journey through those accidents — some famous, some forgotten — not to undermine them, but to reveal the deeper rhythm they were hiding. Every chapter reinterprets a celebrated "quantum miracle" through the UED lens and shows how *coherence, not coincidence*, has always been the author of progress.

So yes, this is a satire — but a respectful one. It laughs at our arrogance but bows to nature's precision. It's a reminder that the universe doesn't reward confusion with prizes. It rewards alignment with revelation.

And sometimes, when we are lucky enough — or quiet enough — we call that revelation a Nobel.

How to Read This Book

This is not a textbook, nor is it a book of mysticism. It is a reconstruction — a journey from confusion to coherence. *The Nobel for Accidents* is meant to be read slowly, reflected deeply, and felt intuitively. You are not merely reading ideas; you are participating in their alignment.

1. Read as a Listener, Not a Learner

Each chapter is a conversation between the field and the mind — between what science discovered accidentally and what existence has always known silently. Do not rush to understand; instead, listen. The rhythm of this book is closer to music than mathematics. Let the meaning unfold like resonance, not instruction.

2. Begin Anywhere, but Return Often

Although the chapters follow a scientific progression — from quantum paradoxes to consciousness — they are written to stand alone as windows into coherence. You may start with *The Tunnel That Wasn't There*, or *The Accidental Human*, or even the *Epilogue*. Each chapter is an entry point into the same continuum. Yet when you finish, circle back. You will find new coherence in what earlier seemed abstract. That is the very nature of this field — understanding refines itself with each alignment.

3. Read in States of Stillness

This book will not argue you into belief. It will *invite you into silence.* Many ideas here reveal themselves best when the mind quiets — early morning, late night, or any moment when you are less rushed by motion. The field responds to attention, not speed. As you read, pause at the aphorisms. They are not decorative; they are anchors. Let each one echo within before you move ahead.

4. Don't Try to Memorize — Try to Recognize

Universal Energy Dynamics (UED) does not ask for faith in equations but for recognition of patterns. When you see words like *ripple*, *alignment*, *tension*, or *stillness*, they refer not to metaphors but to living principles. Notice how they reappear in physics, psychology, biology, and human behavior — that repetition is the proof of coherence.

5. Read Science as Story, and Story as Science

Every discovery retold here — Planck's quantization, Martinis' superconducting qubits, or the myth of the "God Particle" — is presented as a *human confession* rather than a technical chronology. Science, at its purest, is not the study of matter but the journey of mind. Read the experiments as expressions of awareness, and you'll see the unbroken continuity beneath them.

6. Accept That Understanding Evolves

Some sections may feel counterintuitive — especially where UED challenges mainstream physics. That discomfort is natural; it's the echo of your old coherence patterns dissolving. Don't seek immediate agreement. Let understanding emerge the way nature resolves interference — gradually, gracefully, inevitably.

7. Read with a Pen — or with Wonder

Highlight, question, annotate. The margins are where your dialogue with the universe begins. Every realization you note down adds a ripple to the continuum of understanding. But if you prefer, read with nothing — only curiosity. Wonder is a better instrument than intellect when approaching coherence.

8. Remember — You Are Part of the Experiment

This book does not end on its final page. Its real continuation happens in you — in how you perceive, create, and respond to the field around you. When you see patterns connecting, coincidences dissolving, and order emerging where chaos once ruled, that is *The Nobel for Accidents* unfolding in your awareness.

Closing Note

"This is not a book to be mastered; it is a mirror to be listened to. The moment you stop reading it and start hearing it, the field begins to answer."

Prologue – The Age of Lucky Discoveries

History remembers breakthroughs as triumphs of reason. Textbooks say *Eureka!* moments arrive because of genius — as if brilliance and destiny meet at a laboratory bench. But when you peel away the legend and read the scientist's own notes, a more human story emerges — one that begins not with certainty but with surprise. The truth is, most discoveries were found, not understood.

The Paradox of Progress

The story of human progress is less a straight line and more a series of beautiful stumbles. Roentgen discovered X-rays while experimenting with cathode tubes that misbehaved. Alexander Fleming left his lab untidy and returned to find penicillin killing bacteria by accident. Planck reluctantly introduced the quantum constant just to make his equations work — he called it an "act of desperation." Even Einstein, the icon of clarity, once admitted, *"I made up my theories, then I found what they meant."*

Every age of science has been an age of *lucky missteps* — the paradox of progress is that we move forward most rapidly when we don't know what we're looking at.

It's as though the universe hides behind a veil of randomness, waiting for our instruments to fall silent and our assumptions to misfire. And in that fleeting stillness, when logic fails, coherence shows itself.

Accidental Truth — Discovery by Misalignment, Correction by Coherence

In the UED lens, these "accidents" are not random at all. They are moments when the observer's field — mental, emotional, or physical — accidentally synchronizes with the Structured Vacuum Energy (SVE) field of nature. This temporary *coherence alignment* reveals patterns that have always existed but were previously blurred by our own conceptual noise.

So, what we call *discovery* is not invention or creation — it's momentary resonance. A human being, for an instant, aligns with the rhythm of the universe, and something clicks. Then, we rush to write equations, name constants, and publish papers — as if to prove we meant to do it all along.

But the pattern is clear:

Misalignment triggers curiosity. Curiosity leads to attention. Attention brings coherence. Coherence reveals truth.

This is not serendipity — it's the universe using our confusion as an instrument of teaching.

Accidents Are Self-Alignments Momentarily Visible

UED asserts that accidents are field phenomena, not coincidences. They occur when *local misalignments collapse into temporary order*, revealing the coherence hidden beneath chaos. That's why every so-called "accidental discovery" feels both surprising and inevitable — surprising to us, inevitable to the field.

When Fleming's mold killed bacteria, when superconductivity appeared at impossible temperatures, when Josephson junctions started "tunneling" without resistance — these weren't exceptions to physics. They were brief windows of perfect ripple alignment in the structured field of energy.

Humans didn't cause these events. They simply *caught them happening.*

Every Time Humanity Stumbled, Nature Took a Step Forward

That single line captures the essence of scientific evolution. Every stumble — every failed hypothesis, every botched experiment — was the universe gently correcting our direction. Human error became nature's choreography.

And in those moments of unintended precision, we mistook coherence for genius. We handed out medals, built theories, and rewrote textbooks. But the universe never stopped smiling — because while we celebrated our discoveries, it knew they were *its* reminders.

The 2025 Nobel Prize in Physics — for macroscopic quantum tunneling and energy quantization — is only the latest entry in this grand tradition of lucky coherence. Once again, humanity stumbled upon a truth it did not fully understand. Once again, the field revealed itself to those who stopped shaking it. And once again, we called it a miracle instead of a mechanism.

The Age of Lucky Discoveries never ended; it merely changed its instruments. We have moved from glass tubes and pendulums to qubits and superconducting circuits, but the story remains the same. We still find by accident what nature has always known by design.

And perhaps that's the quiet beauty of it all — that the universe keeps letting us call our accidents discoveries, while patiently waiting for us to call them coherence.

Welcome

"The next Nobel will not be for discovery — it will be for coherence."

Chapter 1 – The Quantum Confession

The Birth of Uncertainty and the Death of Continuity

It began not with certainty, but with embarrassment.

At the dawn of the 20th century, physics—so proud of its precision—was quietly falling apart. Equations that once mapped the heavens now broke down in a simple furnace. Energy, that obedient servant of calculus, suddenly refused to flow smoothly. The elegant curves of classical theory shattered into jagged fragments of contradiction.

And in the middle of that intellectual wreckage, a desperate man named **Max Planck** reached for a patch.

He wasn't seeking revolution; he was just trying to make the numbers behave. But when he inserted a small constant—an act he later called *"a desperate trick"*—the universe blinked, and continuity died. Reality, once thought to be an unbroken stream, was suddenly rewritten as a sequence of jumps. Light began to arrive in packets. Energy came in parcels. The cosmos, once fluid, had turned into Morse code.

It was as if nature herself whispered:

"If you insist on measuring motion in fragments, I'll show you a fragmented world."

That whisper became the first confession of modern physics—a quiet admission that we no longer understood what energy *is*. The age of uncertainty had begun.

1.1 The Crisis of Continuity

At the end of the 19th century, science stood at the peak of its own confidence. Physicists believed they had decoded the universe. Newton's laws ran the planets, Maxwell's equations ran the light, and the calculus of smooth change explained everything in between. The world, to them, was a grand mechanical clock — wound by God perhaps, but perfectly predictable thereafter.

Continuity was sacred. Everything flowed. Motion, energy, heat, and time were all smooth fabrics woven together by mathematics. There was no place for jumps, gaps, or uncertainties. Determinism wasn't just a theory — it was a philosophy, a worldview.

And then, suddenly, nature stopped cooperating.

The Blackbody Disobedience

A simple furnace — the blackbody — began misbehaving. Physicists tried to calculate how much energy it should emit at different wavelengths of heat. At long wavelengths, their formulas worked perfectly. But as they moved toward the ultraviolet end of the spectrum, something bizarre happened: the equations predicted infinite energy.

An infinite furnace! A glowing disaster.

The proud equations of classical thermodynamics collapsed under their own logic. Nature refused to follow their smooth mathematics, as if mocking human confidence:

"You want continuity? Try infinity."

The phenomenon became known as the *ultraviolet catastrophe* — but it was not just a technical error. It was the first crack in the illusion of a perfectly continuous universe. For the first time, physics looked into the mirror and saw a paradox staring back.

Equations at the Edge of Reality

Every 19th-century equation—no matter how elegant—broke down at the edges of energy, motion, and light. Newton's laws failed at high speeds, thermodynamics failed at small scales, and Maxwell's waves couldn't explain how light sometimes behaved as particles. The mathematics of continuity had reached its domain of misalignment.

It wasn't the universe that was discontinuous — it was human understanding. The tools that once defined order were suddenly describing chaos. The elegant curves of certainty had turned into cliffs of contradiction.

A Desperate Rescue, Not a Revolution

When Max Planck entered this chaos in 1900, he wasn't looking for glory or revolution. He was simply trying to fix the furnace problem. He adjusted his equations like a surgeon trying to stop a patient from bleeding out. He had no grand theory — just a desperate parameter, a small mathematical constant to make the numbers behave.

That constant, h, would later become one of the most famous in science — but in that moment, it was merely a confession. A quiet acknowledgment that continuity, the great idol of classical physics, had failed.

Planck had not discovered quantum mechanics. He had simply placed a bandage over the wound where determinism had torn. The real meaning of that act — the birth of quantization — would haunt physics for the next century.

Thus began the modern crisis of continuity — the realization that beneath the smooth surface of equations lay something granular, rhythmic, and misunderstood. Not chaos, but structure. Not randomness, but alignment.

The problem wasn't that energy was discontinuous. The problem was that we were measuring it through a discontinuous mind.

1.2 Planck's Desperate Fix

In the winter of 1900, Max Planck faced a mathematical ghost. His equations — elegant, precise, and beautifully classical — were failing to describe the simplest of physical truths: the color of heat. The blackbody curve, that rebellious furnace of physics, refused to obey any known law. Every attempt to reconcile the data with theory led to nonsense — infinite radiation, absurd energy, and the collapse of continuity itself.

For a physicist raised on determinism, this was heresy.

Planck, the disciplined German mind, was no revolutionary. He was a believer in order, reason, and symmetry. Yet he found himself cornered by nature, forced to make a move that went against his own convictions. And so, one evening in October 1900, he quietly inserted a constant — a small, awkward number, later called h, into his equation.

Not to change the world, but to *make the math stop screaming.*

An Act of Desperation

Later, Planck admitted the truth. He called it *"an act of desperation."* He had no philosophical justification, no physical insight. It was, quite literally, a numerical trick to calm the chaos.

He didn't believe energy came in packets. He didn't imagine nature jumping in discrete steps. He only wanted a formula that fit the data — something the experimenters could stop arguing about.

Ironically, that little patch — born of necessity, not vision — became the seed of a revolution. In trying to save continuity, Planck had unknowingly signed its death certificate. He didn't invent quantum mechanics — he *accidentally exposed* it.

The Patch to Preserve Continuity

Textbooks today revere Planck's constant as the cornerstone of modern physics — the birthmark of quantization. But few remember what it really was: a desperate attempt to preserve the old world. Planck was not declaring the universe to be discontinuous; he was desperately trying to make it continuous again.

He treated his constant as a mathematical correction, a calibration for the universe's apparent misbehavior. He was like a composer who, finding the orchestra out of tune, quietly rewrote the score to match the noise.

The world, however, heard it differently. Scientists saw this "correction" as a revelation: that energy itself was granular, jumping between discrete states. Planck's rescue operation had become a revolution — though he himself remained skeptical of it for years.

The UED Translation — What Really Happened

From the lens of Universal Energy Dynamics (UED), Planck's equation was not the birth of discontinuity — it was the first human detection of structured equilibrium.

His mysterious constant, *h*, was not a random quantization factor. It represented the minimum unit of ripple tension in the Structured Vacuum Energy (SVE) — the field that underlies all energy manifestation.

When energy is confined within a boundary (like the walls of a blackbody cavity), the SVE doesn't radiate freely; it forms standing ripple modes — stable harmonics of tension. Each discrete energy level that Planck's formula predicted corresponds to one of these ripple modes reaching equilibrium within the field.

In simpler words:

Planck didn't discover quantization. He discovered ripple coherence, without realizing it.

What appeared to him as *energy in packets* was actually energy in balanced ripples, the field's way of distributing tension in perfect harmony. The apparent discontinuity was a byproduct of our measurement limits — not of nature's design.

Accident #1 — Quantization as Forced Harmony

Thus, the first great "quantum discovery" was not a triumph of insight, but an accidental glimpse of universal order. Planck forced harmony into his mathematics — and stumbled upon the fact that nature already lived in it.

He believed he was inventing a patch to fix misalignment; in truth, he had touched the very structure of alignment itself — the Structured Vacuum Energy field striving for equilibrium through ripple tension.

He had uncovered, unknowingly, the first empirical fingerprint of the universe's coherence — and humanity called it "quantization."

In that one reluctant act, Planck opened the door that would later bewilder Einstein, inspire Bohr, and torment Schrödinger. The quantum world was born — not from revelation, but from desperation. And yet, through that accident, nature had finally begun to reveal her deeper truth: not that energy is broken into pieces, but that coherence, when misunderstood, *looks* like fragments.

1.3 Einstein's Uneasy Acceptance

If Planck's discovery was an act of desperation, Einstein's was an act of discomfort.

In 1905—the miracle year that made him immortal—Albert Einstein proposed something he did not really believe in: that light itself comes in packets. He called them *Lichtquanten*—light quanta. It was a radical idea, but not a joyous one.

Einstein was a man who loved continuity. He worshiped smoothness, symmetry, and causality. In his mind, the universe was a grand, flowing geometry, not a stuttering orchestra of discrete notes. To claim that light—the most continuous thing in nature—arrived in bursts was almost offensive to him.

Yet nature cornered him, just as she had cornered Planck.

Reluctant Revolution

The photoelectric effect refused to behave. When physicists shone light on a metal surface, electrons flew out—not gradually, but instantly, and only when the light exceeded a certain frequency. Classical theory said intensity should matter, not frequency. But the experiments stubbornly disagreed.

Einstein, compelled by evidence rather than belief, extended Planck's desperate idea: he suggested that light energy wasn't spread out like a wave but delivered in discrete lumps, each of energy $E = h\nu$.

He solved the puzzle, yes—but it *felt wrong*. His solution worked because it was mathematically obedient, not because it was conceptually satisfying. Even decades later, Einstein still confessed unease. He wanted unity, not duality; continuity, not collapse. His private letters show a man haunted by the ghost of his own idea— *"I am not satisfied,"* he wrote. *"It seems to me like a temporary expedient."*

He had discovered what worked, but not why it worked.

The Discomfort of Being Right

Ironically, the man who gave quantum theory its first solid footing never truly accepted its foundation. He treated the photon not as a real particle but as a mathematical trick, a bookkeeping device to preserve energy conservation. But the experiments would not forgive him. Every new observation—Compton scattering, photoelectric emission, atomic spectra—seemed to confirm the existence of these indivisible packets of light.

Einstein, the prophet of relativity, was now chained to a theory that violated his deepest conviction: that nature is continuous. He had, in effect, traded coherence for correctness.

The UED Translation — What Einstein Actually Found

From the Universal Energy Dynamics (UED) perspective, Einstein's discomfort was not a flaw; it was intuition resisting misinterpretation. He was right to feel uneasy—because light never travels as particles.

In UED, what Einstein interpreted as *quanta of light* are actually localized ripple resolutions—brief moments when field tension realigns and releases stored coherence. Light is not a stream of pellets flying through space; it is a continuous ripple network, resolving in measurable packets only when coherence breaks locally.

Each "photon" is not a thing, but an event—a temporary tension relief within the Structured Vacuum Energy (SVE) field. When the ripple field is coherent, energy appears wavelike. When coherence collapses locally—say, upon interaction with matter—the same field expresses itself as quantized energy release.

Thus, UED preserves what Einstein longed for: continuity, causality, and determinism—without denying the observed discreteness.

In short:

The photon does not exist; the appearance of one does.

Accident #2 — The Particle Illusion

And so came the second Nobel-scale accident. Einstein, trying to defend energy conservation, inadvertently gave birth to the myth of the photon—the particle of light. It was a clever placeholder that lasted more than a century.

What we now call "wave–particle duality" is simply ripple duality—the same field seen in two different coherence states. Continuous when aligned, quantized when disturbed. The illusion of particles emerges only at the boundaries of broken coherence—where measurement, observation, or interaction introduces asymmetry into the field.

Einstein never meant to divide light into bullets; he simply wanted to make peace between equations and experiments. But that uneasy peace became doctrine. His temporary expedient became physics' permanent paradox.

He once said, *"God does not play dice with the universe."* And in UED's view, he was absolutely right. The dice were never real—only ripples misread as randomness.

Einstein's unease was not his weakness; it was his truth. He had glimpsed the continuity beneath quantization but lacked the framework to name it. UED gives that framework at last.

1.4 The Patchwork Expands

When Planck's constant entered physics, it didn't heal the wound — it just stopped the bleeding. But the patient — classical physics — was still trembling on the table. Continuity had cracked, and every physicist of the new century rushed in with their own bandages.

It was an era of intellectual triage.

Each theorist built a model that solved one problem and created another. The 20th century became the century of patchwork physics — beautiful mathematics draped over unhealed paradoxes.

Bohr's Circling Electrons: The First Bandage

In 1913, Niels Bohr stepped forward with a model that looked like planetary motion redesigned for atoms. Electrons, he claimed, orbited the nucleus in fixed shells, jumping between them only when they absorbed or emitted quanta of energy.

The idea was charming — mechanical enough to feel classical, quantized enough to sound modern. But it was neither truly classical nor truly quantum. Bohr's atom was a diplomatic compromise between truth and tradition — a truce, not a theory.

His quantized orbits "worked," but only for hydrogen. The moment the atom got more complicated, Bohr's neat circles dissolved into guesswork. The patch was holding, but only with duct tape.

Heisenberg's Matrix Maze

Then came Werner Heisenberg, the young radical who decided to stop pretending atoms were anything like tiny solar systems. He threw away the visual picture altogether and replaced it with matrices — abstract grids of numbers that described what electrons did, not what they *were.*

It worked astonishingly well. It predicted spectral lines with surgical precision. But it also severed the last thread connecting physics to intuition. Heisenberg's world didn't just abandon images — it abandoned meaning. You could calculate everything, but understand nothing.

It was as if physics had turned blind but developed super-hearing — sharp enough to recognize nature's tune but not to see the instrument playing it.

Schrödinger's Counterattack — The Wave Dream

In 1926, Erwin Schrödinger couldn't bear it anymore. He wanted a picture, a flow, a rhythm — something continuous. So he turned to waves. If electrons couldn't be particles, maybe they were waves vibrating in invisible fields.

His equation, elegant and lyrical, restored the music of motion. For a brief moment, physics felt continuous again. But only until the question arose: *"Waves of what?"*

No one could answer. The waves weren't in water, air, or ether — they existed in "probability space." Schrödinger had rescued beauty but lost physical reality. The universe sang again, but through a mathematical ghost.

The Language of Survival

Between Bohr's orbits, Heisenberg's matrices, and Schrödinger's waves, a new field called quantum mechanics emerged. It was an extraordinary success — predictive, testable, precise. But it was also a confession:

We don't know what's happening; we just know what happens next.

Quantum mechanics became an emergency language — elegant, powerful, but semantically hollow. It could predict every behavior of nature, yet explain nothing about nature's being. The equations fit, the meaning didn't.

Physics had shifted from understanding to fitting. Nature became an algorithmic puzzle rather than a living continuum. Theorists learned to calculate, not to comprehend.

The Ripple Beneath the Patchwork (UED View)

Under Universal Energy Dynamics, all three models — Bohr's circles, Heisenberg's matrices, and Schrödinger's waves — were describing different projections of the same underlying field behavior. Each was a two-dimensional patch sewn over a four-dimensional fabric — the Structured Vacuum Energy (SVE) — where real motion was not mechanical or probabilistic but ripple-based coherence.

- Bohr glimpsed ripple tension stabilization in circular symmetry.
- Heisenberg translated alignment interactions into numerical transitions.
- Schrödinger sensed field continuity, but lost it to abstract probability.

They each touched the truth but never saw the field itself. The quantum theory that followed was not a theory of reality — it was a dictionary of approximations.

Mathematics Describing Symptoms

And so, the great irony of the 20th century unfolded: as the math became more exact, the meaning became more uncertain. Physicists didn't stop believing in continuity — they stopped needing it. Precision replaced perception; symbols replaced substance.

What had once been the science of cause became the art of correlation. Mathematics, once a window into the universe, became a mirror reflecting our own assumptions.

"Mathematics began describing the symptom, not the structure."

And physics, the grand pursuit of reality, quietly turned into the maintenance of successful illusions.

1.5 Schrödinger's Cat — When Philosophy Lost Its Patience

By the 1930s, quantum mechanics had become an unchallenged empire. It predicted with uncanny precision, dazzled laboratories, and terrified philosophers. But beneath its mathematical brilliance, one problem remained unsolved — *What was actually real?*

Enter Erwin Schrödinger, one of the few who dared to ask the forbidden question. And when he asked it, he didn't use equations. He used a cat.

The Cat Paradox — A Box Full of Confusion

In his thought experiment, Schrödinger placed a cat inside a sealed box with a radioactive atom, a vial of poison, and a trigger mechanism. According to quantum logic, the atom had a 50% chance of decaying, releasing the poison. Until observed, the atom — and by extension, the cat — was said to be in a *superposition* of both states: decayed and not decayed, dead and alive.

It was meant as sarcasm — a philosophical prank to expose the absurdity of taking mathematical formalism too literally. But instead of sparking doubt, it became doctrine. The cat, poor creature, was now the mascot of quantum weirdness.

Schrödinger had intended to embarrass the theory; physicists canonized it instead.

The Triumph of Probability Over Reality

From that point onward, probability replaced mechanism. The universe was no longer said to *do* things; it was said to *choose outcomes* based on statistical odds. Observation became an active ingredient — as though the universe was waiting for human attention to decide whether to exist.

Causation was replaced by correlation; existence became a coin toss until looked at. It was a profound philosophical surrender disguised as sophistication.

The mathematicians were satisfied; the philosophers were exhausted. And so, physics — once the proud heir of logic — quietly outsourced reality to the observer.

The UED Perspective — One Cat, One Field

Under Universal Energy Dynamics (UED), the cat paradox evaporates instantly. There is no superposition of "dead" and "alive" states — because there are no isolated outcomes in a coherent field.

The Structured Vacuum Energy (SVE) field holds both the cat and the atom as part of a single coherent system. The cat isn't split into two possible realities; the *observer's awareness* is split by misalignment in the ripple field.

The act of observation introduces disturbance — a ripple interference that temporarily disrupts coherence. From the observer's side, this looks like "collapse." From the universe's side, nothing collapses at all.

The cat doesn't need saving; the observer does.

When coherence is maintained, reality remains singular and continuous. "Superposition" is simply a misunderstanding of undisturbed alignment — the quiet coexistence of multiple ripple potentials within one stable field.

Reality never divides — only perception flickers between phases of understanding.

When Philosophy Lost Its Patience

Schrödinger's cat became the moment when philosophy finally threw up its hands. It couldn't keep pace with a science that now demanded faith instead of logic. Quantum mechanics had become a religion of equations: *believe the math, not the meaning.*

But in truth, the cat never symbolized uncertainty. It symbolized the arrogance of incomplete perception — the human tendency to mistake its own limitations for nature's complexity.

UED restores simplicity: Reality doesn't split, it stabilizes. Existence doesn't collapse, it continues. The cat, the atom, and the observer are one ripple field — misaligned only in awareness, not in being.

"The cat was never both dead and alive — it was we who couldn't decide which part of the field we were looking at."

1.6 UED Translation — The Ripple Equation

By the time quantum mechanics reached maturity, it had become fluent in a strange dialect — a mathematical language that could predict everything yet explain nothing. The centerpiece of this language was the wavefunction (ψ) — a symbol revered by physicists and feared by philosophers.

To them, ψ was both mystery and miracle. It described the likelihood of where a particle might be, what it might do, or how it might change — but not what it *is*. It was not an equation of motion, but of expectation.

This is where UED steps in and quietly restores meaning.

Probability as Misalignment

In the UED framework, there are no dice rolling inside nature. Probability functions do not represent *mystical uncertainty*; they map how far a system has drifted from perfect coherence.

Each quantum probability distribution is a misalignment map — a chart of how ripple tension varies within the *Structured Vacuum Energy (SVE)* field when constrained by geometry, energy, or boundary conditions.

What looks like "probability density" is simply a measure of local ripple strain — how tension prefers to redistribute in space-time. The peaks of ψ correspond to stable alignment regions; the troughs mark unstable ones.

So when we calculate $|\psi|^2$, we're not finding the "chance" of something existing — we're visualizing how alignment organizes itself under constraint.

"Uncertainty" isn't a law of nature — it's the symptom of unresolved coherence.

The Real Meaning of the Wavefunction

Each wavefunction is therefore a field map — a description of how ripple tension rearranges itself to preserve equilibrium. The equation does not describe a wandering particle but a continuous field of alignment and relaxation, the field's attempt to keep balance amid imposed boundaries.

When you change the potential, you change the field geometry; ψ reshapes accordingly. It's not a new probability — it's a new tension configuration.

What quantum mechanics calls *eigenstates* are actually standing ripple modes, each one representing a coherent tension pattern stabilized by the SVE. These are the same "quantized energy levels" Planck first glimpsed — now recognized as the natural harmonics of structured equilibrium.

Collapse and Superposition — The Language of Misunderstanding

The so-called *collapse of the wavefunction* is not the universe choosing an outcome; it's the re-alignment of ripple tension after interaction. Observation merely disturbs the field's symmetry, forcing it into a new equilibrium that we interpret as a "result."

Likewise, *superposition* is not the coexistence of contradictory realities — it is unresolved coherence within the same ripple field. Multiple potential alignments overlap until an external influence — measurement, boundary, or decoherence — forces one dominant alignment.

No magic. No mystery. Just the field's natural discipline — coherence maintaining itself as long as possible before surrendering to imbalance.

Continuity Restored

UED reveals that continuity never died; only our instruments — and our interpretations — did. Quantum mechanics fragmented the view because its mathematics sampled the field at discrete points, slicing continuous tension into calculable chunks. But the field never stopped being smooth — we just stopped being able to observe it coherently.

Our technology captures misalignment snapshots and calls them *particles*. Our equations describe those fragments and call them *probabilities*. Yet underneath it all, the ripple field flows, unbroken — a still sea misread through a strobe light.

Visual Analogy: The Ripple Spectrum

If one were to sketch it, the picture would look like this:

1. **Classical Curve:** a smooth wave — the ideal of continuity.
2. **Quantum Step:** chopped into discrete energy bands — the illusion of quantization.
3. **Ripple-Mode Spectrum (UED):** structured harmonics of tension, seamlessly continuous but organized in natural nodes.

Where classical physics saw flow, and quantum physics saw fragments, UED sees structured stillness — continuous but organized, rhythmic but unbroken.

"The wave never collapses — only the observer's understanding does."

1.7 The Confession Itself

The deeper quantum theory succeeded, the more its founders doubted it. They could predict the outcome of every experiment —

and yet, they no longer recognized the universe they were describing. Precision had replaced comprehension. The equations sang, but no one understood the lyrics.

And quietly, one by one, the pioneers began to confess.

The Uneasy Fathers of the Quantum Faith

Planck, the reluctant founder, admitted it first. He had called his quantization "a purely formal assumption" — a mathematical convenience, not a physical truth. Even years later, when the world hailed him as the father of the quantum, he sighed, *"I did not invent the quantum theory — I was forced to accept it."*

Then came Einstein, who helped shape the new order but refused to worship it. He saw the numbers work but hated what they implied. "I cannot believe," he wrote, "that God plays dice with the universe." He wasn't being poetic — he was being logical. He simply could not accept that existence itself had become a game of probability.

Bohr, in contrast, embraced the uncertainty but not the meaning. When challenged, he retreated behind riddles: *"No phenomenon is a phenomenon until it is observed."* It sounded profound, but it really meant: *We don't know what's happening until we look, and even then, we might be changing it.*

Schrödinger, horrified by what his own wave equation had unleashed, lamented that physics had turned "indistinct and ghostly." His cat-in-a-box was not a metaphor for curiosity; it was a scream of philosophical despair.

Together, these men had built the temple of quantum mechanics — and then refused to pray in it.

They Knew It Worked, But Not Why

Their dilemma was unique in the history of science. Never before had a theory been so *useful* yet so *incomprehensible*. Every experiment confirmed it; every mind resisted it.

They knew the formulas produced results — but no one could explain what the formulas *meant*. They had decoded the language of nature's behavior but not her grammar. The equations fit like a glove, but no one knew what hand they were covering.

This was the real confession:

They knew the theory worked — but not why.

The genius of the 20th century was not that it explained the world, but that it taught us how to predict it without understanding it. In other words, physics stopped asking *why things happen* and settled for *how often they do*.

UED: The Missing "Why"

Universal Energy Dynamics (UED) provides what the early pioneers never found — the structural reason behind their mysterious success.

Energy appears quantized not because it *jumps*, but because ripple tension locks into discrete standing modes within the Structured Vacuum Energy (SVE) field. Quantization is not a law — it is a pattern of equilibrium. Each "quantum state" is a stable ripple mode where field tension finds harmonic balance, like notes on a stretched string.

When energy transitions between these states, it doesn't leap discontinuously — it realigns. The apparent "quantum jump" is the observer's misinterpretation of a smooth reconfiguration of field geometry.

What Planck and Einstein saw as fragments are simply the nodes of one continuous wave system — the universe balancing itself in discrete harmonics.

Randomness Was Never Random

What they mistook for randomness was unresolved coherence. The probabilistic nature of quantum theory is not an inherent property of reality — it is the statistical shadow cast by our inability to measure full alignment.

In UED, probability distributions vanish when coherence is fully mapped. When the ripple field is aligned, outcomes are deterministic and continuous. When it's misaligned, interference and uncertainty appear — not as laws of nature, but as artifacts of observation.

The universe was never gambling; we were just looking through broken glass.

The Unspoken Truth

If the pioneers of quantum theory could speak again, they might admit what they always suspected — that the universe they described was not strange at all, only misunderstood. They saw fragments of coherence and built an empire on them. They named the fragments particles, probabilities, and paradoxes — but what they were truly observing was the rhythm of structured stillness.

The same rhythm UED calls ripple coherence — the field's eternal effort to remain whole.

"They solved the mystery without understanding the story. The equations worked because the universe was merciful enough to align itself for them — just long enough to call it science."

1.8 The Cost of Losing Continuity

When quantum mechanics became the official language of physics, something subtle but devastating happened — the soul of science began to evaporate.

For centuries, physicists had been guided by a sacred intuition: that nature was continuous, intelligible, and profoundly ordered. Newton's falling apple, Faraday's invisible fields, Maxwell's waves — they all told a story of smooth connectedness. Causality was not just a rule; it was a relationship. Every effect had a cause, every cause a traceable origin, and energy flowed through existence as an unbroken thread of motion and meaning.

Quantum theory severed that thread.

The Triumph of Calculation Over Comprehension

Once probability entered the equations, physics shifted from philosophy to bookkeeping. What mattered now was not *why* nature worked, but *how well* one could predict her next move. Meaning was replaced by method. The new physicists were no longer interpreters of truth; they were technicians of accuracy.

By mid-century, the mantra of the modern physicist had become:

"Shut up and calculate."

It sounded pragmatic — even heroic — but it was actually a confession of defeat. A way to escape the discomfort of not understanding. Generations of brilliant minds were trained to manipulate mathematical ghosts rather than to seek physical coherence. Intuition — once the scientist's compass — was now treated as superstition. Logic became mechanical, and insight became irrelevant.

The universe was no longer a conversation; it was a code to be cracked. And the physicist was no longer a listener — he was a programmer.

The Philosophical Debt

This shift carried a spiritual cost far beyond science. When continuity was dismissed, so was connection. The universe, once seen as an intelligent whole, was fragmented into probabilities without purpose. Energy became statistics, and existence became a random assembly of uncorrelated events.

By cutting off continuity, physics unintentionally divorced energy from intelligence — as though motion could exist without intent, and form without understanding. The harmony between mind and matter, once obvious to philosophers from Heraclitus to Einstein, was now dismissed as poetic indulgence. But poetry had never been the enemy of science — arrogance was.

The irony is striking: In trying to explain everything, physics stopped explaining itself. It lost the very thread that made knowledge meaningful — the awareness that every law of motion is also a law of mind.

The UED Restoration

Universal Energy Dynamics (UED) rebuilds that bridge — quietly, logically, and beautifully.

It restores continuity as the true nature of existence, not as an assumption but as a structural reality. In UED, every form of energy, every wave, every quantum, is a manifestation of Structured Vacuum Energy (SVE) striving for equilibrium — a continuous, living field. There are no gaps between cause and effect, no jumps between levels — only ripple realignments within the same coherent substrate.

UED does not deny the success of quantum mechanics; it completes it. It says:

"Your equations are right — your interpretation is wrong."

Where quantum theory saw randomness, UED sees resolution. Where it saw collapse, UED sees re-alignment. Where it saw uncertainty, UED sees coherence waiting to be understood.

Continuity never left nature — only human perception lost it. And as we restore that perception, we restore not just physics, but philosophy itself.

The True Cost and the True Reward

The cost of losing continuity was the loss of wonder — the quiet understanding that science and spirit were never meant to be enemies. The reward of regaining it will be the reunion of intelligence and energy, of law and life, of the measurable and the meaningful.

For when continuity returns to thought, coherence returns to existence. And when coherence returns, the universe is no longer strange — it is simply whole.

"We did not lose continuity in the universe — we lost it in our minds. UED does not discover it again; it merely remembers."

1.9 "When math loses meaning, nature hides behind probability."

The first Nobel-scale "accident" — quantization — began as a patch of desperation and ended as a pillar of dogma. What started as Planck's reluctant mathematical fix turned into a full-fledged belief system — one that mistook necessity for nature.

Over the years, the bandage hardened into doctrine. Physicists began to worship the very paradoxes that once embarrassed their founders. Randomness became sacred, uncertainty became wisdom, and discontinuity became law. The mystery was no longer something to be solved — it was something to be preserved.

It is a strange irony of history: Science, which once sought simplicity, began defending complexity. And thus, what should have remained an *accidental success* became the intellectual religion of the 20th century.

The Cycle of Repetition

But the pattern did not end with Planck or Einstein. Each "quantum miracle" that followed repeated the same choreography:

- A paradox emerged.
- A clever equation patched it.
- Nature complied for a while.
- The meaning was quietly lost.

From tunneling to superconductivity, from entanglement to decoherence, every so-called discovery followed this rhythm of misalignment — a misunderstanding dressed as progress.

At every turn, coherence was mistaken for randomness, alignment for chance, structure for uncertainty. The field was whispering the same truth in a thousand different languages, but we kept responding with mathematics instead of understanding.

UED's Quiet Correction

Universal Energy Dynamics (UED) does not reject the mathematics — it decodes it. It tells us that quantization was never proof of discontinuity; it was evidence of structured harmony. The universe wasn't breaking itself into pieces — it was organizing its stillness into modes of balance.

Where quantum mechanics saw "mystery," UED sees mechanism. Where it found probability, UED finds pattern. And where it declared collapse, UED reveals re-alignment.

What the pioneers glimpsed but could not describe, UED articulates with clarity: that all energy, all motion, all interaction, are manifestations of one continuous ripple field seeking coherence.

From Confusion to Coherence

Thus ends the first act of misunderstanding — the birth of uncertainty and the death of continuity. But the story does not stop here. The next chapters will revisit the same stage, where each "quantum miracle" will be stripped of its mystique and seen for what it truly is: an accidental glimpse of alignment.

The cat survived, the equations didn't. What perished in quantum theory was not the particle or the wave — it was meaning itself. And now, meaning must return.

"The cat survived, the math didn't. From here onward, every mystery will be another confession of coherence."

Chapter 2 – The Tunnel That Wasn't There

The 2025 Nobel and the Myth of Quantum Tunneling

Stockholm, October 2025. The world of physics erupts again — another "quantum miracle." The Nobel Committee has just announced its winners: **John Martinis, John Clarke, and Michel Devoret**, awarded for "demonstrating macroscopic quantum tunneling and energy quantization in superconducting circuits."

The headlines read like poetry written by confusion.

"Particles Pass Through Barriers." "Quantum Effects at Human Scale." "The Impossible, Proven Again."

The world applauds, but nature, if she could speak, might have sighed.

Because nothing actually passed through anything. No particle leaped, no energy escaped, no miracle occurred. Something far simpler — and far deeper — took place: *the field found stillness.*

For over a century, physicists have called "tunneling" one of nature's strangest tricks — a particle appearing where it shouldn't, defying classical logic, sneaking through forbidden zones. But under **Universal Energy Dynamics (UED)**, there is no trick, no paradox, and no tunnel. There is only *ripple re-alignment* — coherence restoring itself through the Structured Vacuum Energy (SVE) field.

When alignment returns, the illusion of impossibility disappears. The supposed "jump" is simply the field relaxing to its next stable mode — like a wave quietly finding its rhythm after chaos.

They cooled their circuits to near absolute zero, silencing every disturbance — and in that silence, the universe revealed her oldest law:

Stillness doesn't cross barriers. It erases them.

That, in truth, was the 2025 Nobel discovery — not a triumph of quantum weirdness, but a confirmation of coherence. They called it tunneling. UED calls it *remembering*.

2.1 The Triumph of the 2025 Nobel

In 2025, the Nobel Prize in Physics went to three brilliant experimentalists — John Martinis, John Clarke, and Michel Devoret — for what was hailed as a landmark achievement: the observation of macroscopic quantum tunneling and energy quantization in superconducting circuits.

To the global physics community, it was nothing short of poetic — a long-awaited confirmation that quantum effects could scale up from the atomic to the visible, from the microscopic to the macroscopic. Nature, it seemed, had allowed humanity a peek behind her last curtain.

The Triumph of the Unthinkable

For decades, quantum mechanics had confined its magic to the invisible — to the world of atoms, electrons, and wavefunctions that no human eye could directly witness. But these experiments brought the unseeable into circuitry the size of a thumbnail. The circuits behaved like particles — *jumping, tunneling, resonating* — in perfect obedience to quantum equations.

The headlines arrived like fireworks:

- "Quantum Behavior Seen at Human Scale!"
- "Superconducting Circuits Defy Classical Physics!"
- "Particles Tunnel Through Barriers — Macroscopically!"

The public was dazzled. The physicists congratulated each other. And the Nobel Committee, faithful to tradition, rewarded mystery over meaning.

Once again, an *accidental coherence* had been crowned as deliberate discovery.

The Official Story

The laureates had cooled their superconducting circuits to near absolute zero, eliminating all thermal noise — the eternal enemy of coherence. They observed a current that appeared to "tunnel" through an insulating barrier between two superconductors. This was seen as the macroscopic equivalent of an electron slipping through a potential wall — but now, instead of a single particle, the *entire current* behaved like a quantum entity.

The data were elegant: energy levels quantized, transitions discrete, probabilities measured. Every instrument agreed. The equations matched. It was physics at its most precise — and yet, at its most paradoxical.

The Myth of Motion

The conventional explanation was simple enough to sound profound: A particle trapped in a potential well sometimes "appears" on the other side — without crossing the barrier, without sufficient energy to climb over it. It simply *tunnels through*, as if the universe occasionally forgets its own rules.

That story had worked for electrons, protons, and even alpha particles. Now it was working for electric currents. Physics celebrated this as proof of "quantum weirdness made visible."

But from the UED perspective, there was no miracle here — only misunderstanding. Nothing tunneled, nothing passed, nothing broke through.

The UED View — When Coherence Pretends to Move

Under Universal Energy Dynamics, what Martinis, Clarke, and Devoret observed was not *motion through a barrier*, but ripple re-alignment within the Structured Vacuum Energy (SVE) field. When the superconducting system reached extreme stillness, thermal agitation vanished, and ripple tension began to behave coherently. At that point, energy did not flow — it *re-phased*. The system didn't "cross" the barrier; the barrier lost distinction.

In other words, the "tunneling" was not an act of defiance against physics — it was physics returning to its natural state of coherence.

The Real Achievement

The true beauty of the 2025 Nobel was not that humanity made particles tunnel through walls, but that it accidentally created the conditions for stillness. When interference is removed, coherence emerges. When coherence emerges, nature stops looking paradoxical. The scientists didn't force nature to behave quantum mechanically — they simply stopped disturbing her.

It wasn't a triumph of technology. It was a revelation of what happens when the field is left alone.

The Central Thesis

The entire myth of "quantum tunneling" collapses under one clear realization:

They didn't break through; they aligned through.

The field was never blocked. The barrier was never real. What they saw was not matter violating a wall — but the Structured Vacuum Energy field correcting a misalignment. It was coherence revealing itself under the guise of an experiment.

A Nobel, once again, for an accident of alignment.

2.2 The Standard Explanation — Tunneling as Escape

For more than a century, quantum tunneling has been described as one of nature's strangest tricks — the phenomenon that breaks all classical logic yet works perfectly in equations. It's taught in every physics classroom as the ultimate badge of quantum weirdness: a particle, trapped inside a potential barrier it cannot possibly cross, somehow appears on the other side.

It doesn't climb, roll, or smash through the wall — it simply *vanishes here* and *reappears there.* No path, no force, no trace. And every generation of physicists has been told to accept it as *proof* that reality doesn't need to make sense.

The Classical Setup

Imagine a small particle — say, an electron — trapped inside a potential well. On one side lies a barrier higher than its energy. Classically, it should bounce back forever, confined by the limits of its own strength. But quantum theory paints a different picture: the wavefunction describing that particle doesn't stop at the barrier — it extends through it, tapering exponentially, and re-emerges faintly on the other side.

The mathematics says the particle has a small but non-zero probability of being found beyond the wall. It's as though the particle were a ghost that sometimes forgets which side of the grave it's buried on.

The textbooks call this *tunneling*. The labs confirm it with real effects — from alpha decay to Josephson junctions. The world applauds, and the paradox deepens.

Mathematics Versus Intuition

The equations of Schrödinger describe this perfectly. The potential barrier is just another curve in the energy landscape; the wavefunction flows through it as if it were mist.

But the human mind, trained in cause and effect, rebels: How can something without enough energy *move* through a region it shouldn't reach? Where does the missing energy come from? And if the particle is "real," how does it survive being in two places at once?

Quantum mechanics offers no satisfying answer — only the comfort of prediction. It says, "Don't ask how. Just accept that it happens." Thus, physics quietly replaced mechanism with mathematics, and paradox became pedagogy.

We were told that uncertainty is not ignorance — it's the nature of reality itself. But what if uncertainty is only our distorted reflection of a deeper order we've stopped looking for?

The Contradiction Hidden in Plain Sight

In classical physics, movement requires energy. In quantum physics, tunneling requires imagination. The equations suggest that something happens — but what exactly happens, no one dares to say.

If energy is conserved, there is no surplus to cross the barrier. If causality is respected, there is no path to follow. If reality is continuous, the "jump" has no meaning. And yet, every experiment affirms that the effect is real.

The contradiction is no longer between theory and nature — it's between understanding and obedience. We were taught to worship probability because we could no longer explain process. And in that surrender, intuition died quietly.

Transition

But perhaps the truth was never paradoxical at all. Perhaps nothing *moved* — because nothing *had to.*

"It never did."

The next section reveals what really happens when energy appears to pass through the impossible — not as escape, but as alignment. Not as defiance, but as resolution.

2.3 The Misleading Miracle

Tunneling was never just another discovery — it became the poster child of quantum weirdness. It was the moment physics declared, almost proudly, that reality no longer needed to make sense. If a particle could appear where it had no right to be, then everything was fair game: uncertainty, duality, probability, paradox — all crowned as the new rulers of science.

Physicists called it "quantum strangeness." Philosophers called it "the death of causality." But nature, as always, stayed silent — because she knew she hadn't done anything strange at all.

From Mystery to Icon

Tunneling quickly became the icon of modern physics — the *proof* that intuition could no longer be trusted. Here was the ultimate badge of quantum faith: an event that could be calculated but not imagined. Particles didn't need permission anymore; they simply appeared where they weren't supposed to.

In classrooms, tunneling was introduced with reverence, as if it were a sacred paradox:

"A particle with insufficient energy crosses a barrier anyway." No one dared to ask *how*. The explanation always ended with the same sentence — whispered like a hymn: "That's just quantum mechanics."

And thus, confusion became canon.

Turning the Mystery Into Utility

While philosophers despaired, engineers got busy. They stopped worrying about meaning and started building devices.

- Transistors exploited tunneling to create switches for the digital age.
- Josephson junctions used it to transmit current across barriers — the foundation of superconducting circuits.
- Quantum bits (qubits) used it to maintain dual states, giving rise to quantum computing itself.

The mystery, once maddening, became profitable. Physics might not have understood tunneling, but technology didn't care — it only needed it to *work*.

And work it did. From the electronics in your phone to the logic gates in supercomputers, tunneling became the invisible backbone of modern civilization. The paradox had found its paycheck.

But even as it built the digital world, the conceptual void behind it remained untouched. What was actually moving? What *crossed* the barrier? How did energy that shouldn't exist still produce an effect that undeniably does?

The answer never came — because the question itself was wrong.

The Assumption: "Particles Pass Through Walls"

The entire foundation of the tunneling story rests on one mental image: that of a particle *passing* through a wall. It's the picture every student draws, every professor describes, and every textbook repeats. The idea is emotionally satisfying — something hidden, struggling, somehow *escaping*. But it's also completely false.

No particle ever crosses a barrier. No wall is ever breached. In reality, there are no "sides" at all. There is only a field, and what we call a "barrier" is merely a region of different alignment within it.

The particle doesn't pass through — the field reconfigures itself to a new equilibrium. From our narrow, time-sliced perspective, this looks like motion or transfer. From the universe's perspective, it's just ripple tension redistributing itself across the same continuum.

To put it plainly:

What physicists call tunneling, nature calls relaxation.

The Reality: Fields Reconfigure Themselves

Under Universal Energy Dynamics (UED), tunneling is not a transport event — it's a tension event. Energy doesn't move; equilibrium shifts. The Structured Vacuum Energy (SVE) field adjusts its ripple tension until both sides of the "barrier" reach a stable configuration. The effect looks like passage, but it's simply the field remembering its coherence.

The circuit doesn't send electrons across; it synchronizes field phases. The current doesn't flow; the ripple field realigns. The barrier doesn't vanish; it becomes irrelevant.

This is why the effect seems instantaneous — there's no travel time for something that never moves. It's not superluminal, it's *subluminal* — below motion, beneath causality, existing at the level of structured stillness.

When the noise stops, the field stops resisting itself. That is tunneling. Not an escape, but a homecoming.

"What physicists call tunneling, nature calls relaxation."

2.4 UED Reinterpretation — The Ripple Doesn't Tunnel

Under Universal Energy Dynamics (UED), the phenomenon of quantum tunneling is not motion, not mystery, and certainly not magic. It is realignment — the reconfiguration of ripple tension within the Structured Vacuum Energy (SVE) field.

What classical physics interprets as a "barrier," and what quantum mechanics describes as "probability penetration," UED sees simply as a localized region of misalignment — a ripple field temporarily out of phase with its surroundings.

When coherence is restored, tension re-equilibrates — and what appears to be a particle "passing through" is actually the field reorganizing itself into a new stable configuration. No particle moves. No wall is crossed. No rule is broken. The illusion of motion is merely the artifact of an observer measuring misalignment through time.

Energy Doesn't Travel — It Re-Phases

In the UED view, energy is not a traveler wandering through empty space; it is the expression of field tension seeking balance. The Structured Vacuum Energy field is everywhere, always. When part of it becomes misaligned — through charge, confinement, or potential difference — ripple tension forms to correct it.

These ripples are not objects; they are patterns of pressure in the field. When one region reaches coherence and another doesn't, tension begins to re-phase between them, not as a flow but as a synchronization of alignment states. This synchronization is what we mistake for movement.

To the measuring instrument, it looks as though energy "leapt" across a gap. But in reality, both sides of the gap simply became phase-matched — like two strings vibrating into harmony through resonance. The barrier never blocked anything; it merely delayed alignment.

The Barrier That Wasn't There

In classical physics, a barrier is an obstacle. In quantum mechanics, it becomes a probability wall — something you can't cross but might mysteriously appear beyond. In UED, it's neither of these things.

A barrier is simply a ripple mismatch zone — a place where local tension gradients distort the natural coherence of the field. It's not a wall in space, but a misaligned state within the Structured Vacuum Energy continuum.

When coherence strengthens — through cooling, phase isolation, or field stabilization — the mismatch dissolves. The field relaxes, tension redistributes, and the barrier ceases to have physical meaning. To the observer, this looks like the particle "tunneling" through, but from the field's point of view, nothing ever moved.

The Apparent Discontinuity

The so-called discontinuity of tunneling — that instant when something "appears" on the other side — is not a physical event. It's an observational illusion, born of trying to measure continuity through instruments that depend on time-based sampling.

Our detectors see discrete before-and-after states but miss the continuous re-phasing that connects them. To an observer bound by time, re-alignment looks like a jump. But in reality, ripple tension is never discontinuous — it's the observer's coherence that collapses, not the field's.

Every "quantum jump," every "wavefunction collapse," every "tunnel event" is merely our fragmented measurement of a field in the process of self-correction.

The True UED Translation

UED replaces the mythology of quantum tunneling with a simple structural principle:

- **Barrier** → Ripple mismatch zone in the Structured Vacuum Energy (SVE) field.
- **Tunneling** → Ripple re-synchronization and coherence restoration.
- **Probability** → Degree of local misalignment (how far from perfect coherence the field is).
- **Transmission** → Field reaching phase continuity across regions.

Thus, the "transmission probability" of a tunneling event is nothing mystical; it's the mathematical measure of how successfully coherence propagates through ripple re-phasing. It is not chance — it is structural order rediscovering itself.

Visual Mapping

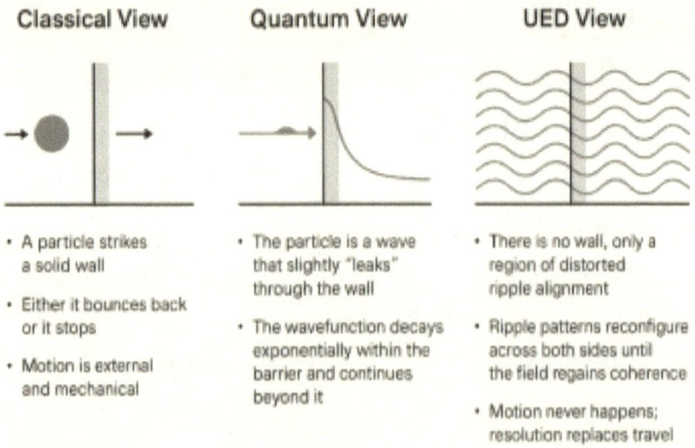

Diagram Explains:

Classical View:

- A particle strikes a solid wall.
- Either it bounces back or it stops.
- Motion is external and mechanical.

Quantum View:

- The particle is a wave that slightly "leaks" through the wall.
- The wavefunction decays exponentially within the barrier and continues beyond it.
- Motion becomes probabilistic.

UED View:

- There is no wall, only a region of distorted ripple alignment.
- Ripple patterns reconfigure across both sides until the field

regains coherence.

- Motion never happens; resolution replaces travel.

Ripple Coherence — The True Mechanism

When the ripple field aligns perfectly, tension gradients drop to zero — no forces, no flows, no losses. This is the zero-resistance state observed in superconductivity, the same condition that made the 2025 Nobel possible. What we call tunneling is not the breaking of limits; it's the return to natural alignment.

In every experiment, whenever noise, temperature, or disturbance is minimized, coherence rises — and "quantum miracles" appear. UED identifies this not as violation but revelation: nature behaving as she always does when left undisturbed.

"Barrier = ripple mismatch; tunneling = ripple re-synchronization. The ripple doesn't pass through — it passes beyond misunderstanding."

2.5 The Experiment Without a Mystery

Every great scientific revelation begins with a lab that gets too quiet. The 2025 Nobel-winning experiments were no exception. Deep within cryogenic chambers, surrounded by the hum of precision electronics and the silence of near-absolute zero, the teams of Martinis, Clarke, and Devoret achieved what physicists once called impossible — macroscopic quantum tunneling and energy quantization in superconducting circuits.

The achievement was monumental — not because nature did something new, but because for the first time in decades, humans stopped interfering long enough to let her be herself.

The Setup: A Playground of Stillness

At the heart of the experiment were superconducting qubits — small loops of current interrupted by Josephson junctions, ultra-thin insulating barriers that separate two superconductors. These junctions were engineered to act as potential wells for electric current — the same way atomic orbitals trap electrons.

According to the standard narrative, the current inside these circuits "tunnels" through the barrier when the system reaches a specific energy threshold, creating quantized states that can be manipulated like artificial atoms. But under the UED lens, nothing "jumped" or "passed through."

What actually happened was far simpler and far more elegant: ripple coherence emerged — not by design, but by silence.

Cooling to Stillness

To make the effect visible, the experimenters cooled their systems to a few millikelvins above absolute zero — temperatures so low that thermal vibrations nearly stopped. Every atom, every electron, every ripple of motion slowed into alignment. The chaotic jostling of thermal energy — the field's most persistent noise — fell silent.

At that level of quiet, the Structured Vacuum Energy (SVE) began to express its natural coherence. Ripple tension, normally distorted by random motion and temperature-induced misalignment, found the freedom to reorganize itself. Energy didn't move — it simply redistributed equilibrium throughout the field.

The supposed "tunneling" wasn't motion — it was resolution.

What Actually Happened Physically

From the UED perspective, here's what unfolded inside those superconducting circuits:

1. **Thermal agitation dropped to near zero**. The chaotic misalignments that normally break coherence disappeared.
2. **Ripple tension stabilized.** The SVE field reorganized its internal structure, forming continuous, low-resistance alignment paths through the circuit.
3. **Macroscopic coherence emerged**. Entire regions of the field — trillions of atoms — began oscillating as one coherent entity.
4. **The Josephson junction "barrier" dissolved.** It didn't vanish physically; it simply lost its distinction at the field level. Ripple tension equalized across both sides, creating a seamless energy landscape.

To an external observer, it appeared as though current had tunneled through an insulator. In truth, nothing crossed. The field simply stopped recognizing the separation.

No Leap, No Travel — Just Re-Alignment

There were no particles leaping, no electrons teleporting, and no energy disobeying classical rules. There was only the continuous re-alignment of ripple tension in a coherent field.

When the field aligns perfectly, resistance ceases to exist — and what appears to us as motion is really restoring symmetry. That's why superconductivity, superfluidity, and tunneling are always discovered in low-noise environments: they are not new phenomena — they are the universe returning to her natural state of stillness.

Why the Experimenters Succeeded

The success of Martinis, Clarke, and Devoret was not a triumph of manipulation but of *minimization*. They succeeded because they removed interference.

- They eliminated temperature, letting the field cool into coherence.
- They eliminated vibration, preventing ripple scattering.
- They reduced electromagnetic noise, allowing the field to stabilize.

And then, when they had finally stopped forcing, fixing, and fighting — nature completed the experiment for them.

"They silenced the noise — and the field spoke."

It was not invention. It was revelation.

Accidental Coherence

Their triumph wasn't in discovering a new phenomenon, but in accidentally recreating the ancient condition of universal balance — the same principle governing all coherence, from atomic orbitals to cosmic equilibrium.

The 2025 Nobel didn't prove that quantum effects can scale up. It proved that coherence is scale-independent. When alignment returns, size becomes irrelevant.

Once again, humanity stumbled upon stillness — and called it a miracle.

2.6 Why Coherence Looks Like Magic

Every age of science has its favorite illusion. For Newton, it was absolute space; for Einstein, spacetime curvature; and for the quantum world, it became instantaneity — the idea that something could happen *without happening in time.*

When misalignment vanishes and the field achieves coherence, events appear to occur instantly. A transition that seems to require time — a particle crossing a barrier, an electron jumping or a current flipping state — suddenly unfolds faster than any causal process can explain. To classical reasoning, it looks supernatural. To quantum formalism, it's "non-local." To UED, it's simply stillness misunderstood as speed.

The Classical Fear of the Instant

Classical physics has always been uneasy with instant change. Causality demands time — a sequence, a before and after. If something happens without transition, the framework collapses.

So when experiments began showing electrons "appearing" across barriers with no measurable delay, scientists panicked — and then celebrated. They called it a quantum miracle, proof that the microscopic world obeys rules beyond logic. But that celebration was premature, because the logic never failed — the interpretation did.

Physics mistook absence of delay for defiance of causality. But an absence of delay only means one thing: the system is not moving *through* time at all — it's reconfiguring beyond it.

The UED Translation — Reconfiguration, Not Motion

In Universal Energy Dynamics, coherence is not speed — it's timeless resolution. The Structured Vacuum Energy (SVE) field operates as a single, continuous whole. When one region of the field realigns, the tension everywhere adjusts *simultaneously* — not through transmission, but through instant equilibrium of a continuous structure.

It's not faster-than-light or supernatural; it's non-local elasticity. Like stretching a single sheet — when you pull one corner, the whole fabric adjusts, not because information "travels," but because the sheet was never divided to begin with.

Thus, what quantum mechanics calls "instantaneous tunneling" or "non-local correlation" is simply ripple coherence acting within the same undivided field.

Nothing jumps. Nothing crosses. The field merely remembers itself.

Alignment Beyond Time

Ripple coherence does not propagate; it equilibrates. It's not an exchange between points in space, but a restoration of uniform tension across the entire structure. This is why coherence acts faster than any causal mechanism — because it doesn't *travel*, it *resolves*.

When the field becomes perfectly coherent, time itself loses meaning. Causality requires sequence; coherence exists as simultaneity. To the observer measuring through time, it looks like a magical instant — a violation of physics. To the field, it's simply timeless continuity.

In tunneling, the barrier doesn't fall, nor does energy move through it. The misalignment between both sides dissolves as equilibrium is restored. The field's ripple tension re-synchronizes, and in that timeless act, both sides of the barrier realize they were never truly separate.

When Stillness Becomes Speed

When motion ceases, and resistance disappears, the only thing left is resolution. From the human perspective, that resolution looks like infinite speed — but it's not speed at all. It's absence of delay — the return to perfect coherence.

This is why every "instantaneous" phenomenon in physics — from superconductivity to entanglement to tunneling — occurs only when noise vanishes. When the field stops fighting itself, time stops measuring. Stillness begins to look like sorcery.

"What we call instant is simply eternity caught in the act of alignment."

The Dissolving Barrier

So, in tunneling, there is no barrier to cross. The "wall" only exists as long as misalignment does. When equilibrium returns, the barrier *dissolves into coherence.* To call that event "tunneling" is like calling peace a form of warfare — it's a contradiction made sacred by habit.

The universe doesn't leap from one state to another. It reconfigures itself continuously until distinction disappears. In that stillness, nothing moves — because everything is already where it belongs.

"Instantaneity is not speed — it is silence remembering itself."
"In tunneling, no barrier is crossed. The barrier dissolves in equilibrium."

2.7 The Myth of Quantum Weirdness

If there is one word that defines modern physics, it isn't *understanding* — it's *weirdness.* Ever since the early 20th century, the scientific community has treated confusion as a badge of honor. What began as intellectual humility became a culture of paradox — a theater where mystery itself is worshipped as truth.

Physicists don't just study phenomena anymore; they perform them. And the audience, dazzled by uncertainty, applauds the strangeness.

The Cult of the Counterintuitive

Quantum mechanics has survived for over a century not because it explains reality, but because it *defies* it so beautifully. The language of paradox gives scientists the thrill of being on the edge of understanding without ever having to step beyond it. The stranger it sounds, the more profound it feels.

They call it "quantum strangeness," "spooky action," "wave–particle duality." These phrases have become poetic shields against simplicity — intellectual fireworks distracting from the plainness of order beneath.

Why this obsession with paradox? Because paradox protects prestige. If something is simple, it becomes accessible. If it's accessible, it's no longer sacred.

So when coherence — nature's most ordinary principle — appears in front of them, they call it "weird." Because admitting it's normal would collapse a century of romantic uncertainty.

Why Physicists Keep Calling Coherence 'Strangeness'

Coherence, under UED, is the natural tendency of the field to return to alignment. But to modern physics, this behavior looks supernatural. Why? Because the human mind, conditioned by motion and causality, cannot imagine order without activity.

When energy reorganizes without travel, when equilibrium restores without exchange, when information synchronizes without transmission — we panic. We assume something "spooky" must be happening. But the spookiness is not in nature; it's in our inability to recognize stillness as a form of intelligence.

What physicists call *quantum weirdness* is simply nature behaving coherently — something we rarely do ourselves.

The Drama of Paradox

Paradox keeps physics exciting. Without it, textbooks would read like meditation manuals — silent, balanced, calm. And calm doesn't sell.

So the community chose drama over clarity. Every unexplained harmony became a headline of defiance:

- Electrons "teleporting" (entanglement).
- Particles "being in two places" (superposition).

- Currents "passing through barriers" (tunneling).

Each discovery added a new costume to the same actor: coherence. And every time she entered the stage, we called her by a different name.

We forgot that the universe isn't performing tricks — it's simply maintaining balance. The weirdness is a reflection of our noise, not her nature.

UED Perspective — The Familiar Stillness

From the Universal Energy Dynamics viewpoint, "weirdness" is not a property of the universe but of the observer's misalignment. When your perception operates within distortion, coherence appears strange. But when the observer aligns, all strangeness dissolves.

In UED, there are no miracles, no paradoxes, and no supernatural exceptions. There is only one rule: ripple alignment. Every phenomenon, from gravity to light, from tunneling to entanglement, is a variation of the same process — the field seeking equilibrium.

Coherence is not quantum; it is universal. The fact that we call it weird only reveals how far our instruments — and our minds — have drifted from natural order.

Every "Quantum Miracle" Is Just Ripple Alignment

Look closely at the sacred list of quantum wonders:

- **Tunneling** → Ripple re-synchronization across misaligned tension.
- **Entanglement** → Coherent field behavior across multiple observation points.
- **Superposition** → Ripple tension unresolved between possible alignments.

- **Collapse** → Resolution into stable equilibrium through interaction.

Different names, same story. Different effects, same cause. Different experiments, same principle: the field seeking stillness.

Nature never needed a new physics; she only needed a quieter audience.

The True Irony

The universe is not weird. It is infinitely consistent. It is the observer who is noisy, fragmented, and impatient — expecting movement where there is only continuity, and surprise where there is only structure.

We built a philosophy of wonder around our own misunderstanding and called it science. But the truth is simpler, calmer, more beautiful:

"Every time nature stabilizes herself, humans call it non-intuitive."

"Weirdness is just unfamiliar stillness." "When nature stops struggling, we start speculating."

2.8 Philosophical Reflection — Stillness as Power

At first glance, the 2025 Nobel Prize was about superconductors, quantum circuits, and microscopic equations. But beneath the scientific jargon, something far greater was revealed — a quiet philosophical truth that modern physics had forgotten: stillness is not weakness; it is power in perfect balance.

The experiment wasn't a victory of human intellect over nature. It was the moment nature reminded us — gently but unmistakably — that all struggle is a symptom of misalignment.

The Victory of Stillness Over Struggle

For centuries, we have defined energy by its aggression — the power to move, to collide, to overcome. Every force was imagined as an act of conquest: energy breaking barriers, matter defying resistance, light escaping darkness.

But the 2025 Nobel experiment inverted that logic completely. There was no collision, no push, no breakthrough. When the scientists cooled their system close to absolute zero, motion ceased, and what emerged was not failure — but perfection. In stillness, the field aligned. In silence, coherence appeared.

Energy didn't conquer the barrier — it stopped needing it. And the moment struggle vanished, nature did what she always does when left unopposed — she restored harmony.

That was not just an experiment. It was a confession: that power has never been the ability to force, but the wisdom to align.

The Deep Analogy — Beyond the Wall

There's an ancient metaphor hidden in this physics:

"The moment you stop fighting the wall, you find yourself on the other side."

It's the same truth found in meditation, in art, in enlightenment — and now, in superconducting circuits. The "wall" we imagine in science, life, or mind exists only as long as resistance does. When coherence returns, the wall dissolves — not because it was broken, but because it was never truly there.

In UED, this isn't poetry — it's physics. A barrier is not an obstacle but a misaligned ripple. When field tension resolves, distinction disappears, and what once felt like separation becomes continuity.

The universe doesn't need to tunnel; it simply stops dividing itself.

The True Message of the 2025 Nobel

What Martinis, Clarke, and Devoret really discovered was not a new quantum effect but an old universal law — that coherence is nature's original state, temporarily disturbed by noise, motion, and misunderstanding. When interference ends, coherence reappears automatically. It doesn't need to be engineered; it only needs to be allowed.

UED translates this revelation into law:

"Energy does not move toward order; it returns to it."

All so-called "quantum phenomena" are not exceptions to physics — they are reminders of its foundation. Every paradox, from tunneling to entanglement, is simply the universe healing itself from fragmentation.

In that light, the 2025 Nobel becomes not just a celebration of experimental success, but a cosmic allegory — the moment when science rediscovered what philosophy and consciousness had always known: that the highest motion is stillness, and the greatest force is harmony.

Closing Reflection

If physics has been a long struggle to understand movement, UED offers a new frontier — the science of alignment. The 2025 Nobel didn't prove that humans had mastered the quantum world; it proved that nature still rewards silence over struggle. Every time we quiet interference, coherence reveals itself. And every time we observe without force, the universe aligns.

The so-called "miracle" of tunneling was never about crossing barriers — it was about remembering we were never separated in the first place.

"Stillness is not the absence of power — it is power perfected. When motion stops, truth moves."

2.9 The UED Equation of Coherence Transfer

After the philosophical clarity of stillness, it's time to give that stillness its mathematics. UED expresses every apparent motion, jump, or tunneling event as nothing more than a transfer of coherence — a structural re-alignment of ripple tension in the Structured Vacuum Energy (SVE) field. Unlike conventional quantum mechanics, this description requires no probabilistic leap, no imaginary barrier, and no uncertainty.

When coherence is perfect, nothing moves, because everything is already aligned.

The Equilibrium Condition

At the heart of UED lies a simple symbolic truth:

$$\Delta T = 0$$

Where ΔT represents the *difference in ripple tension* between two adjacent field regions. When $\Delta T \rightarrow 0$, the system reaches perfect equilibrium, and no real tunneling occurs — because there is no imbalance left to correct.

In classical terms, this would mean "no energy transfer." But under UED, it means the entire field has re-phased into coherence. Energy isn't lost or moved; it has settled into uniform stillness.

When $\Delta T \neq 0$, local misalignments emerge — what physicists interpret as potential barriers or wells. These are not "objects" but gradients of tension within the SVE substrate.

Ripple Phase Realignment (RPR)

All tunneling events, all so-called "quantum jumps," are governed by what UED calls Ripple Phase Realignment (RPR):

$$RPR: \quad E(x,t) = E_0\, f(\phi)$$

where

- $E(x,t)$ is the local field energy density,
- E_0 is the equilibrium field baseline, and
- $f(\phi)$ describes the *phase relation* of local ripple alignment.

A change in ϕ (phase) without change in E_0 represents a pure re-phasing — a shift in *how* the field stores tension, not *where* it moves.

Thus, during a "tunneling" event, the field reconfigures phase continuity across a region of misalignment, rather than transporting any particle across space.

Barrier Potential as Field Misalignment

In quantum notation, a potential barrier $V(x)$ is defined as a region of higher energy. In UED translation, it is simply a misalignment potential, denoted:

$$\Delta\Phi = f(\Delta T)$$

where

- Φ represents the phase-coherence potential of the SVE field,
- $\Delta\Phi$ quantifies how far a region is from natural alignment, and
- ΔT links that misalignment to measurable tension.

The larger the $\Delta\Phi$, the stronger the distortion in local ripple coherence — the "thicker" the imaginary barrier. But once $\Delta T \to 0$ through RPR, $\Delta\Phi$ vanishes — and so does the concept of a wall.

From Probability to Coherence Ratio

Quantum theory interprets tunneling probability as a statistical chance of a particle's presence beyond the barrier. UED replaces this with a coherence ratio, expressing how efficiently a region's ripple field re-synchronizes with its surroundings:

$$P_{UED} \propto C = \frac{\text{Aligned Ripple Density}}{\text{Total Ripple Density}}$$

Where C = 1 corresponds to total coherence (barrier dissolved) and C < 1 indicates partial misalignment (barrier persists). Hence, what appears as "probability" is actually a measure of coherence, not randomness.

No event is uncertain; only the observer's alignment is.

The Real Mechanism

1. **Before Alignment:** Misaligned ripple zones create local gradients ($\Delta T > 0$, $\Delta\Phi > 0$). Classical models mistake this for a confining potential.
2. **During Alignment:** RPR activates — phase continuity begins re-establishing across the region. Observers detect a "tunneling probability wave."
3. **After Alignment:** $\Delta T \to 0$, $\Delta\Phi \to 0$. The field stabilizes; coherence becomes complete. To the observer, a particle has "appeared" beyond the barrier.

But in truth, nothing crossed — the field simply remembered its symmetry.

Conceptual Diagram

The ripple pattern before tunneling shows distorted, disordered wavefronts. After tunneling, the ripple field appears denser and more uniform — not because something passed through, but because tension equalized.

View	Representation	Interpretation
Classical	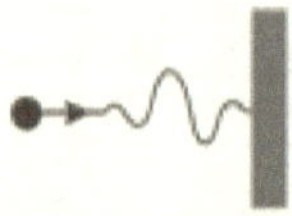	Motion confined; barrier absolute
Quantum	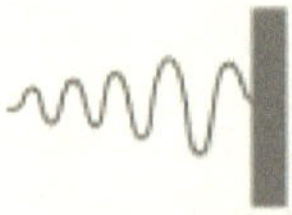	Probability decay within potential
UED		Misalignment dissolving through phase coherence

Unified Principle

The UED Equation of Coherence Transfer may be summarized symbolically as:

$$\boxed{\Delta T \to 0 \Rightarrow \Delta\Phi \to 0 \Rightarrow RPR \to C \to 1}$$

When ripple tension difference vanishes, phase misalignment vanishes, and the field regains perfect coherence. That's when "tunneling" ends — not by motion, but by dissolution of distinction.

"Tunneling ends not when something crosses, but when nothing needs to." "The field doesn't jump; it heals."

2.10 Boundaries Never Truly Existed

Every discovery tells two stories: the one science celebrates, and the one nature actually reveals. The 2025 Nobel Prize — awarded for "macroscopic quantum tunneling" — was officially about particles, currents, and barriers. But underneath that language, something far deeper was happening. The event was not about crossing boundaries at all. It was about revealing that boundaries never truly existed.

The 2025 Nobel as a Case Study in Accidental Alignment

The experimenters believed they had forced the quantum world into visibility — coaxed subatomic behavior to appear at human scale. But in truth, they had simply stepped out of the way.

By cooling matter to the edge of absolute stillness, they silenced interference long enough for the field to remember itself. Their apparatus didn't create coherence; it merely stopped disturbing it.

In that instant, the Structured Vacuum Energy (SVE) field — unbroken, undivided, omnipresent — expressed its natural order. What they interpreted as "tunneling through a barrier" was simply ripple coherence returning to equilibrium. A discovery born not of conquest, but of surrender.

It was, quite literally, an accident of alignment — nature teaching us again that truth reveals itself when resistance ends.

The Discovery Was Never About Crossing Boundaries

Humanity loves the romance of breakthroughs — the idea of crossing walls, defying limits, breaking rules. But the universe operates on a quieter law: nothing needs to be broken when everything is already connected.

The so-called "barrier" in the superconducting circuit was never an obstacle; it was a misalignment. When coherence returned, the distinction dissolved. The current didn't move *through* the barrier — the field stopped recognizing it.

The event we glorified as a miracle of motion was actually the disappearance of separation. Nature didn't cross anything — she simply became whole again.

Reframing the Miracle

We called it tunneling because we measured it as movement. But movement is the illusion of the impatient observer. Under UED, the same event is the restoration of continuity — a glimpse of how the universe behaves when not fragmented by perception.

So, the 2025 Nobel wasn't proof of quantum strangeness. It was proof of universal sanity. A reminder that coherence is the natural state, and that all "miracles" are simply returns to that equilibrium.

In UED, nothing violates physics — physics only violates its own understanding when it forgets that stillness is the foundation of everything.

Closing Aphorism

"They didn't break through; they aligned through."

A statement that turns the entire narrative of discovery upside down — from action to awareness, from force to harmony. It's not the story of matter doing the impossible, but of intelligence finally recognizing the possible.

The Next Question — and the Next Chapter

If cooling brought stillness to matter — restoring coherence at the smallest scales — could the same order explain how the universe holds itself together at the largest ones?

If ripples align within atoms, what happens when ripples align across galaxies? Perhaps the same coherence that dissolves a microscopic barrier is also what bends the fabric of existence.

"Where motion ends, the field bends — and we call it gravity."

With that, the stage shifts from the quantum to the cosmic — from the silent circuits of superconductivity to the curved continuity of spacetime itself.

- "Every time coherence is achieved, humanity names it a miracle."
- "Tunneling isn't motion — it's misunderstanding."
- "Stillness doesn't pass through barriers; it makes barriers irrelevant."

Interlude — Between Confession and Coherence

The first chapter was an admission; the second, a revelation.

In *The Quantum Confession*, we saw the moment when physics lost continuity — when humanity mistook mathematical adjustment for truth. Planck patched a failing equation, Einstein tolerated discomfort, and the world of smooth causality shattered into fragments of probability. That was the beginning of *quantum confusion*: when equations began predicting perfectly, but explaining nothing. We learned to calculate, not to understand. We replaced insight with syntax — and called it progress.

A century later, in *The Tunnel That Wasn't There*, the same story replayed itself — this time, with better instruments and colder laboratories. Yet beneath all the noise of precision, nature repeated the same quiet lesson: *nothing actually happens when alignment is restored.* The Nobel in 2025 didn't honor discovery; it honored rediscovery — the reappearance of stillness beneath the chaos of our curiosity. What we celebrated as tunneling was simply continuity refusing to die.

Between these two moments — Planck's desperation and Martinis's silence — lies the hidden rhythm of modern science: we break what already works, and when nature heals herself, we call it a breakthrough.

The quantum pioneers stumbled upon discontinuity because they looked for energy in isolation. The modern experimenters stumbled upon coherence because they finally stopped looking altogether. The first tore reality apart; the second briefly let it mend. And both times, the universe whispered the same word: *alignment.*

UED stands in the space between these centuries — not as rebellion, but as reconciliation. It does not rewrite quantum mechanics; it completes its unfinished sentence. It tells us that energy never came in packets; it came in patterns. That probability was never randomness; it was ripple memory. And that tunneling was never movement; it was simply *the field remembering itself through stillness.*

So, between confession and coherence stands the real story of physics — the slow awakening from fragmentation to flow, from probability to pattern, from noise to the natural silence of understanding.

Chapter 3 – The Cold Truth

When Matter Remembers Itself

It begins with silence — the kind that deepens, not the kind that ends. In the heart of every cooling experiment, there's a moment when noise stops being random and starts being meaningful. The hum fades, vibrations vanish, and temperature, that most human of measures, begins to betray its own illusion.

We say the system "loses heat." We imagine atoms slowing down, motion fading, energy draining into emptiness. But that's not what really happens. What truly happens is that *matter begins to remember.*

As the last traces of agitation dissolve, the Structured Vacuum Energy field — the quiet fabric beneath all matter — reclaims its balance. Every particle, every wave, every ripple stops arguing about its direction. Chaos ceases to be a form of activity and becomes what it always was: *a form of forgetfulness.*

At high temperatures, the universe forgets itself in motion. At low temperatures, it remembers — not through stillness, but through perfect synchronization. This is why superconductors awaken when others think they're dying. Why superfluids flow endlessly when motion should have ceased. And why, near absolute zero, the laws of physics start whispering instead of shouting.

Temperature, it turns out, was never the measure of motion. It was the measure of *misalignment*. When heat disappears, what remains is not absence — it's order.

"At absolute zero, nothing freezes — everything listens."

That is the cold truth. And through that truth, UED reveals a quiet revolution: what we once called cooling is simply the universe remembering its own coherence — the great act of stillness by which all energy, all matter, and all life return to their original harmony.

3.1 The Temperature Illusion — Measuring Motion, Misunderstanding Meaning

Temperature has always been one of humanity's favorite illusions — a number we trust so deeply that we've stopped questioning what it really measures. From the first thermometers to modern cryogenic labs, we've believed that temperature tells us how much energy a body has, how fast its atoms move, how alive its matter is. But what if that belief itself is incomplete?

Classical physics defines temperature as the average kinetic energy of particles in a system. More motion means more heat. Less motion means more cold. Simple, measurable, convenient — but misleading. Because what temperature truly reflects is not motion, but misalignment.

The Paradox of Cooling

When a system cools, we think it's losing energy. We imagine particles slowing down, vibrating less, fading into stillness. But if energy were truly vanishing, how could extraordinary order suddenly emerge at those low temperatures?

The paradox is undeniable: the closer we move toward absolute zero, the more organized matter becomes. Resistance disappears. Friction ceases. Electrons pair, atoms unify, and flows become eternal. Cooling — which by all classical logic should bring death — instead brings coherence.

The supposed silence of cold is not emptiness; it's structure returning. Every reduction in temperature is not a loss, but a refinement — a system rediscovering harmony after a long argument with itself.

When Motion Fades, Memory Returns

Consider the great miracles of low-temperature physics:

- **Superconductivity**, where electrical resistance vanishes.
- **Superfluidity**, where liquids flow endlessly without losing energy.
- **Bose–Einstein condensation**, where atoms merge into a single collective state.

Each of these is born not from new energy, but from the end of noise. As random motion falls, the field stops fighting itself. The ripples in the Structured Vacuum Energy (SVE) substrate realign, forming one seamless rhythm across the entire domain. The field doesn't die — it remembers.

At higher temperatures, every particle vibrates out of sync, each trying to maintain its individuality. Cooling doesn't punish that individuality; it simply reveals that individuality was never separate — just misaligned.

UED Reinterpretation — Energy Never Leaves, It Realigns

In Universal Energy Dynamics (UED), temperature is not the amount of energy present but the degree of internal disagreement within the field. The hotter the system, the higher the ripple distortion; the colder the system, the greater the alignment. When temperature drops, energy doesn't leave — it reorganizes. The field's tension, once chaotic, finds its coherent structure again.

In UED terms:

"Cooling doesn't remove energy; it restores alignment."

The total energy of the system remains constant — only its pattern changes. What we perceive as heat leaving the body is really the system surrendering its internal argument and falling back into resonance with its surroundings.

A Simple Analogy

Imagine an orchestra out of tune — every musician playing slightly off-pitch, slightly off-time. The sound is loud, chaotic, and heated. Now imagine the conductor lowering the volume, slowing the tempo, and letting every instrument listen. Suddenly, the chaos collapses into harmony. The music becomes quiet — yet far more powerful.

That is what cooling does to matter. You don't stop the orchestra by lowering the volume; you let it tune.

The Beginning of a Deeper Realization

Temperature, then, is not a measure of energy; it is a measure of misunderstanding. The thermometer does not tell us how alive matter is — only how misaligned it currently feels. The colder a system gets, the less it argues with itself. And when all argument ends, it achieves the highest possible state of awareness: perfect coherence.

The "cold truth" is not about frozen lifelessness — it's about clarity. In that clarity, matter stops pretending to move and begins remembering what it is made of: stillness structured by intelligence.

3.2 What We Call Heat — Misalignment Density

Heat has always been treated as motion — a restless dance of particles, each vibrating, colliding, transferring energy from one to another. It is the story we've been told since the dawn of thermodynamics: when atoms move faster, the system becomes hotter; when they slow down, it cools. And yet, this description, though measurable, has always been incomplete. It tells us what happens, but not why it happens.

Heat as Random Motion — The Old Paradigm

In classical physics, heat is disorder. Each particle moves on its own, without reference to others — collisions are chaotic, vibrations incoherent, and the total motion, though immense, is meaningless. There is energy, yes, but no purpose. The sum of random vectors, endlessly changing direction, achieves nothing but agitation.

To us, this looks alive. But from the universe's perspective, it's confusion.

UED's Reframing — Heat as Ripple Misalignment

Under Universal Energy Dynamics (UED), motion is not the cause of heat — it is the symptom of misalignment. Every atom, every molecule, every structure is an expression of ripple tension within the Structured Vacuum Energy (SVE) field. When those ripples are well aligned, the field is calm, coherent, and stable. When they fall out of phase — when tension gradients emerge — the system begins to churn. This internal struggle is what we experience as heat.

Thus, temperature is not a measure of kinetic energy; it is a measure of misalignment density — the concentration of ripple disagreement within a field.

$$T \propto \rho_{\text{misalignment}}$$

Where T represents temperature and $\rho_{\text{misalignment}}$ represents the local density of incoherent ripple tension.

Heat as the Density of Distortion

The hotter a system becomes, the more disordered its internal ripples grow. The field that binds its particles loses phase coherence. Vibrations grow sharper but less synchronized, collisions more frequent but less meaningful. This is why heat always feels like resistance — because resistance is nothing but misalignment amplified.

Every furnace, every flame, every fever is the sound of the field arguing with itself.

In contrast, when a system cools, its ripple distortions begin to fade. Ripples fall back into mutual rhythm; field tension redistributes evenly. The energy doesn't vanish — it stops conflicting. Cooling, therefore, does not reduce energy; it reduces contradiction.

The SVE Perspective — Energy Never Fights Itself

The Structured Vacuum Energy field is fundamentally neutral. It doesn't generate chaos on its own — chaos is what emerges when localized distortions prevent the field from expressing its equilibrium. Heat, then, is not an entity or substance; it is the observable signature of lost coherence.

This view transforms the entire language of thermodynamics. Temperature ceases to be a scalar of speed and becomes a scalar of stability. The higher the temperature, the greater the distortion in the field; the lower the temperature, the closer the field is to remembering its natural rhythm.

In this sense, heat is not "stored energy" but structural tension. And temperature is simply how loudly that tension vibrates.

Cooling as Reconciliation

When a system cools, it doesn't die — it reconciles. Each ripple slows its rebellion, each vibration re-enters harmony, and the chaotic energy of motion transforms into the silent energy of coherence. This is why superconductors, superfluids, and coherent condensates arise in cold environments: the field finally stops fighting.

Cold is not absence — it is alignment. It is the field remembering its shape, its origin, its purpose.

Philosophical Reflection

We once thought heat meant vitality and cold meant death. But the deeper we go into matter, the more we realize the opposite. When the field is noisy, it forgets itself. When it becomes quiet, it knows.

So perhaps it's time to retire an old myth — that heat is life and cold is void. In truth, heat is confusion made visible, and cold is the wisdom of matter remembering itself.

"Heat is not motion — it's misunderstanding made physical."

3.3 Cooling and the Return to Memory

Cooling is often imagined as decay — a slowing down, a fading away of life and motion. But beneath that popular metaphor lies an extraordinary reversal: as matter cools, it doesn't lose energy; it recovers awareness. The deeper the temperature falls, the closer matter comes to remembering its original structure — the pure alignment from which it was born.

Memory as Structural Coherence

In Universal Energy Dynamics (UED), *memory* is not stored in neurons, circuits, or codes. It is an inherent property of the field — the natural tendency of ripples to restore coherence when disturbance ceases. Every atom, every pattern, every system in the universe carries within it the echo of perfect order. The act of cooling doesn't awaken something new — it simply allows this order to surface again.

As the temperature drops, the random misalignments — once seen as heat — begin to vanish. Ripples that were colliding incoherently start resonating together, regaining a shared rhythm. This synchronized behavior is memory in action — the field recalling its own geometry.

To "remember" in this sense means to *re-member* — to reassemble what was once fragmented. Cooling, then, is not suppression; it is integration.

Why Superconductors and Superfluids Emerge Near Absolute Zero

When matter cools close to absolute zero, it begins to exhibit behaviors that defy conventional explanation — superconductivity and superfluidity. In these states, energy moves without loss, without friction, without effort. The reason, under UED, is simple: the Structured Vacuum Energy (SVE) field regains near-perfect alignment.

At high temperatures, the field is noisy — ripple misalignments scatter tension and dissipate energy as resistance. But as temperature decreases, the field becomes quiet. Its internal contradictions collapse into harmony. Every ripple becomes aware of every other, forming one continuous network of self-consistent tension. In that harmony, nothing resists. Energy doesn't need to "flow" — it simply exists coherently.

This is why, near absolute zero, matter achieves its most "magical" behaviors:

- **Superconductivity** — charge ripples align so completely that resistance, the measure of misalignment, disappears.
- **Superfluidity** — atomic ripples synchronize so thoroughly that the concept of viscosity loses meaning.

These are not anomalies; they are *restorations* — the universe briefly returning to its natural equilibrium.

The "Frozen" State Isn't Dead — It's Listening

In conventional thermodynamics, we imagine the frozen state as the end of all motion, the death of energy. But under UED, stillness is not silence — it is *listening.* At near-zero temperatures, every ripple becomes perfectly sensitive to every other. Noise is gone, distortion is gone, separation is gone — what remains is pure receptivity.

The system is no longer acting; it is aware. Its ripples vibrate in such subtle synchrony that the distinction between one and many fades. In that equilibrium, the field does not move — it *understands.*

Cold is not the opposite of heat; it is its correction. When all interference is removed, the SVE field stops trying to speak — because it is finally heard.

UED Aphorism

"At absolute zero, nothing freezes — everything listens."

This is the most profound truth of temperature. Cold is not emptiness; it is clarity. It is the moment the field, after endless motion and noise, sits in perfect stillness and remembers what it has always been — a single, coherent intelligence manifesting as matter.

Philosophical Reflection

We call it freezing because we mistake silence for death. But in truth, it is the opposite — the resurrection of meaning within matter. When the field stops struggling, its structure becomes audible. Superconductivity and superfluidity are not mechanical achievements — they are moments of awareness in the material world.

To cool a system is to make it listen — to return it to the memory it never truly lost. And in that listening lies the most powerful kind of intelligence: the intelligence of harmony.

3.4 Superconductivity — The Alignment of Charge Ripples

Superconductivity has long been hailed as one of nature's most spectacular miracles — a state where electricity flows without resistance, seemingly liberated from the rules of ordinary matter. Discovered by accident in 1911 and explained decades later through quantum pairing models, it has always carried an air of mystery: why does perfect conduction arise only when everything slows down?

Under Universal Energy Dynamics (UED), the answer is not hidden in complex mathematics or exotic materials. It lies in something simpler, more universal, and profoundly more elegant: alignment.

The Classical View — Electrons in Pairs

Conventional theory, through the Bardeen–Cooper–Schrieffer (BCS) model, tells us that superconductivity occurs when electrons form delicate pairs known as Cooper pairs. These pairs move in perfect synchronization through the atomic lattice, avoiding the usual scattering that creates electrical resistance. The mathematics works beautifully, and the model predicts critical temperatures and properties with admirable precision.

But even its founders knew — the mechanism is descriptive, not explanatory. Why should electrons, the most repulsive entities in physics, suddenly begin to cooperate? What invisible conductor turns chaos into choreography?

The BCS theory identifies the harmony but not its origin. It describes the tune, not the tuning.

The UED View — Ripple Tension, Not Pairing

In the UED interpretation, there are no individual electrons traveling through the lattice — no particles forming couples. There is only the Structured Vacuum Energy (SVE) field expressing charge as ripple tension. What we call "current" is not a flow of particles; it is the propagation of alignment across this tensioned field.

When a system cools below a certain threshold, the misalignment density — the degree of internal ripple disorder — drops sharply. The individual oscillations of charge cease to compete. Their tension fields synchronize into one coherent standing pattern.

At that moment, resistance disappears not because electrons move freely, but because there is no relative motion left to resist. Every ripple vibrates in phase with every other. Friction, by definition, requires difference — and difference no longer exists.

The material, once noisy, has remembered how to listen. It becomes a single, continuous ripple domain — an orchestra in resonance with itself.

Misalignment and Resistance

Resistance is nothing more than the friction of disagreement — the cost of internal conflict within the field. When neighboring ripples are out of phase, energy is wasted in correction, generating heat and distortion. But when ripples align perfectly, every fluctuation reinforces the whole.

UED expresses this as a field principle:

$$R \propto \rho_{\text{misalignment}}$$

Where R is resistance and $\rho_{\text{misalignment}}$ t is the local density of phase disorder. As $\rho_{\text{misalignment}} \to 0$, the field achieves total phase coherence. Current, then, is not a stream of charge — it is ripple continuity. Superconductivity marks the moment when charge stops being movement and becomes memory.

Zero Resistance — The Signature of Coherence

When total phase coherence is reached, $\Delta T = 0$, meaning there are no remaining ripple tension gradients. Every region of the field vibrates in identical phase — one continuous rhythm of existence. At this stage, energy no longer "flows" because it doesn't need to move; it simply exists everywhere in harmony.

This is why superconductivity can maintain a current indefinitely — not because energy is trapped in motion, but because the field has stabilized into a state where motion is no longer required for persistence.

Superconductivity, then, is not about electrons escaping resistance — it is about the field escaping misalignment.

The "Quantum Effect" as Field Memory

What we call a "quantum effect" is not something alien to classical physics; it's classical order restored at its most fundamental level. When the system becomes silent enough, the Structured Vacuum Energy field remembers its own continuity. This is why every "quantum" state — from tunneling to superfluidity — arises only under conditions of minimal interference.

Superconductivity is therefore the material manifestation of universal remembrance — matter regaining access to its built-in intelligence through stillness.

The laboratory measures it as zero resistance; UED reads it as perfect recollection.

Philosophical Reflection

If heat is noise, and resistance is disagreement, then superconductivity is forgiveness — the reconciliation of all misalignments. The material, once fragmented into countless competing vibrations, falls silent and becomes one coherent field.

The "quantum leap" was never a jump forward in energy. It was a return inward to harmony.

"The so-called quantum effect is simply the field's memory restored through silence."

3.5 Superfluidity — The Flow That Isn't Flowing

Few sights in physics are as hauntingly beautiful as liquid helium below the lambda point. Pour it into a bowl, and it refuses to stay put. It climbs walls, crawls up and over the rim, and spills itself across the surface — not as droplets, but as a living skin of liquid that seems to ignore gravity. Spin it in a cup, and it never stops rotating. Try to make it still, and it defies you with eternal motion.

This is superfluidity — the state where viscosity, the most inevitable drag of the material world, vanishes completely. For the classical physicist, it is one of nature's strangest spectacles. For UED, it is one of her most honest.

Helium's Miracle — The Behavior Below the Lambda Point

Helium becomes a superfluid at 2.17 K, the famous lambda point, named after the shape of its specific heat curve. Below that threshold, the liquid stops behaving like a collection of atoms and starts behaving like a single organism.

- It climbs the walls of its container.
- It flows through microscopic pores that no other liquid can penetrate.
- It sustains motion indefinitely without friction or decay.

Classical physics explains this through quantum mechanics: the atoms collapse into the same quantum ground state — a Bose–Einstein condensate. In this state, all particles share the same wavefunction, behaving as one collective entity.

This explanation works mathematically. But under Universal Energy Dynamics (UED), we no longer need to imagine invisible quantum identities. We can see the same event as the ripple field restoring complete coherence — the entire atomic field vibrating as one.

The UED Perspective — Phase Continuity, Not Flow

In the UED framework, superfluidity is not "flow" in the classical sense. No particles are racing through space, and no energy is being carried across the liquid. Instead, the Structured Vacuum Energy (SVE) field underlying the atoms achieves near-perfect ripple alignment.

When alignment becomes global, phase continuity extends across the entire liquid. Each atom's ripple merges with the next, and together they form one seamless field of tension — a single continuous wave of existence.

In this state, "motion" ceases to have meaning. There is no beginning or end to flow, no push or pull, no direction or resistance. The superfluid doesn't move; it remains coherent across change.

To the observer, it looks like motion without loss. But to the field itself, it is simply stillness expressed through continuity.

Motion Without Loss — Stillness in Motion

In all ordinary matter, motion comes at a cost — friction, resistance, heat. These are not natural consequences of motion but of misalignment. When the field within a substance is disordered, motion produces turbulence, and energy dissipates as noise.

In a superfluid, however, misalignment density collapses toward zero. Ripple tension equalizes across the entire volume. The field doesn't have to move to propagate — it simply remains phase-coherent everywhere.

This is why a spinning cup of superfluid helium never stops turning: the system has no internal friction to consume its motion. What appears to be perpetual rotation is actually unbroken resonance — a system existing in complete internal agreement.

Superfluidity, therefore, is not about motion defying resistance — it is about stillness so complete that motion and rest become indistinguishable.

The Parallel With Superconductivity

Superfluidity and superconductivity are not two different miracles — they are two versions of the same event. Both arise when misalignment density approaches zero and the ripple field achieves domain-wide coherence.

- In superconductivity, the charge field aligns — electrical ripples synchronize.
- In superfluidity, the atomic field aligns — mass ripples

synchronize.

In both cases, the SVE field beneath the system remembers its original order. The phenomenon is not "quantum" at all; it is universal coherence made visible. When the field stops struggling, both charge and mass behave as aspects of the same stillness.

Superconductivity is the music of electrons remembering harmony; superfluidity is the dance of atoms remembering stillness.

The Field's Memory of Oneness

At extreme cold, the SVE field enters a state of awareness — not consciousness in the human sense, but coherence in the structural sense. Every ripple is aware of every other ripple's phase; every disturbance is instantly compensated by the field's self-consistent tension. No signal travels — the system simply remains itself.

This is why UED refers to superfluidity as the return of memory. It is the material world's brief remembrance of its true nature: a continuous, tensioned field where energy, matter, and motion are only different expressions of equilibrium seeking itself.

Analogy — Existing Together

Imagine a thousand dancers moving across a stage. When each dances alone, chaos rules — collisions, noise, resistance. But when they move in perfect rhythm, when no step is out of phase, they don't appear to move at all — they appear to glide, a seamless continuity of grace.

That's what superfluid helium is doing. Not flowing, not moving — simply existing together, perfectly.

"Superfluids don't flow — they continue existing together."

Philosophical Reflection

Superfluidity is not the escape from matter's limitations — it is matter rediscovering that it never had any. When the field listens deeply enough to itself, it no longer needs to move to connect. It simply *is*. And in that state, even motion becomes meditation.

3.6 The Myth of Freezing — Stillness as the Ultimate Activity

For centuries, humanity has been haunted by a single idea: that cold means death. From early thermodynamics to modern cryogenics, we've been taught that absolute zero — 0 Kelvin — is the point where motion ceases, energy fades, and the universe falls silent. But under the lens of Universal Energy Dynamics (UED), that silence is not extinction — it's enlightenment.

The great myth of freezing is not that it kills activity, but that it hides a deeper kind of activity we have never learned to see.

The Conventional View — The End of Motion

According to classical physics, as a system cools, the motion of its constituent particles slows. At 0 Kelvin, all kinetic motion should cease — atoms frozen in perfect stillness, incapable of vibration or interaction. This is the so-called "ground state of matter" — the ultimate void of activity.

It's a comforting simplification, but an incomplete one. Because even when all measurable motion ends, energy does not vanish. The field beneath — the Structured Vacuum Energy (SVE) — remains alive, alert, and in equilibrium. To say motion stops is to mistake silence for absence.

The UED Correction — Stillness, Not Stagnation

UED redefines this apparent nothingness. At absolute zero, mechanical motion — that is, misaligned, kinetic agitation — indeed stops. But field activity persists as structured tension within the SVE continuum. The universe does not stop existing simply because we've run out of heat to measure.

The SVE field is not dependent on temperature; it *is* the baseline of temperature. What we call "zero" is simply the moment when the field stops conflicting with itself. All distortions, vibrations, and disagreements dissolve into perfect coherence. The universe doesn't freeze — it listens.

The Field Never Freezes — It Harmonizes

Freezing, in the UED sense, is not immobilization but resolution. The field ceases to express itself through motion because it no longer needs to. All ripple tensions — once pushing, pulling, oscillating — are now balanced in perfect reciprocal alignment.

Every point in the field mirrors every other. There are no gradients to trigger energy transfer, no phase delays to create waves. What remains is a self-sustaining, self-aware equilibrium — a timeless structure in which existence continues without struggle.

To the classical observer, that looks like nothing. To the UED observer, that is everything functioning perfectly.

Even in Deep Stillness, Tension Persists

The Structured Vacuum Energy field is never idle. Even in perfect equilibrium, ripple tension persists — not as motion, but as the silent architecture of existence. It is this tension that maintains universal balance: the equilibrium between potential and expression, stillness and structure, existence and perception.

At absolute zero, this tension is no longer dynamic; it is resolved. It holds everything in coherent suspension — the ultimate poised state where energy neither flows nor fades. It is the resting heartbeat of reality, vibrating not in sound but in awareness.

The Field at Absolute Zero — The Perfect Listener

Imagine the universe as a great orchestra. At high temperatures, every instrument plays its own melody — fast, loud, chaotic. As temperature drops, the sound softens; the players begin to listen. At absolute zero, no one plays at all — yet the music hasn't stopped. The silence itself has become the symphony.

That is what the SVE field is at 0 Kelvin — a perfect listener. It receives everything, interferes with nothing. It is coherent, receptive, and aware — not in a conscious or psychological sense, but as a structural intelligence that sustains the harmony of being.

Stillness, therefore, is not the absence of activity but the completion of purpose. It is when the field ceases to *seek* equilibrium because it *is* equilibrium.

The Conceptual Inversion

To grasp the truth of cold, we must invert our cultural metaphors. The UED framework replaces the traditional thermodynamic hierarchy with a more accurate one:

Concept	Classical View	UED Reinterpretation
Heat	Activity, vitality, energy	Noise, distortion, internal disagreement
Cold	Death, stillness, emptiness	Awareness, coherence, alignment
Zero	Absence of energy	Unity of all tension — equilibrium realized

Thus, the journey toward zero is not a descent into void, but an ascent into oneness. Heat is not life — it is confusion. Cold is not death — it is clarity.

Stillness as the Ultimate Activity

The irony of existence is that the more still something becomes, the more complete its activity actually is. Motion is a symptom of imbalance; stillness is the resolution of all motion. To reach perfect rest is to participate in the most profound form of activity possible — activity so complete, so coherent, that no further change is necessary.

At absolute zero, matter remembers everything it ever was, because nothing in it contradicts itself anymore. It has achieved the one state the universe forever seeks: unity without conflict, energy without movement, power without noise.

- "The universe never freezes — it only stops arguing."
- "Stillness is not death; it's energy with nothing left to prove."
- "Zero is not the end of motion — it's motion resolved." "Heat is noise; cold is awareness; zero is unity."

3.7 The Law of Thermal Alignment

Every civilization that studied energy eventually met its philosophical wall at the same point — the Second Law of Thermodynamics. It was the law that seemed to say: *no matter what you do, order will always decay.* Entropy will increase, systems will lose structure, and the universe will drown in randomness. For over a century, this law has been taken as the ultimate truth — the mathematical equivalent of death.

But under Universal Energy Dynamics (UED), the Second Law is not wrong — it's incomplete. It describes what happens when coherence is observed through a narrow, local lens. It does not account for the larger field — the universal continuum that never loses its balance.

The Law of Thermal Alignment restores that missing dimension.

The Classical Second Law — A Tragedy of Context

The Second Law states that in an isolated system, entropy (disorder) always increases. Energy disperses, temperatures equalize, and order dissolves into equilibrium. This is why heat flows from hot to cold — from regions of higher energy density to lower — until everything becomes uniform.

But the traditional interpretation carries a hidden assumption: that equilibrium is the *death* of order, not its perfection.

Thermodynamics, in its human form, measures activity and calls it life. It measures stillness and calls it decay. It never asked: *What if equilibrium is not the loss of structure, but the restoration of coherence?*

UED Reversal — The Field Always Evolves Toward Alignment

Under UED, entropy is not the march toward disorder — it is the visible side of an invisible process: the universe resolving misalignment. Locally, things appear to decay; globally, they reconnect.

Every time a system radiates heat, what it's really doing is shedding ripple noise — discharging its internal misalignment into the greater field, where the energy re-enters coherence. This is why heat naturally flows outward: not to die, but to rejoin alignment.

Thus, UED replaces the Second Law's premise of inevitable disorder with a deeper principle:

$$\frac{dC}{dt} \geq 0$$

Where C represents *coherence* — the degree of alignment between ripple tensions within the field. This means that, over time, coherence always increases, though not always locally or visibly.

Entropy, as defined by human observation, may appear to increase — but only because the observer is measuring local chaos rather than global order.

Entropy Reframed — The Shadow of Coherence

In classical physics, entropy is the measure of randomness or the number of possible microscopic states consistent with a macroscopic condition. But in UED, entropy is reinterpreted as the measure of ripple imbalance — the extent to which a field's internal tension deviates from coherent equilibrium.

The higher the misalignment density ($\rho_{misalignment}$ the higher the entropy. Cooling reduces this misalignment, bringing ripples into synchronization — and thus, reducing entropy in that local domain.

However, globally, coherence still increases. Even as one system cools and another warms, the total field finds new balance through redistribution of tension. No information is lost — only reorganized.

Entropy, then, is not the universe running down; it is the universe balancing itself.

The Law of Consequences (LOC) and Thermal Behavior

All thermodynamic processes — heating, cooling, dissipation — are governed by the Law of Consequences (LOC), UED's fundamental behavioral law of energy. LOC states that every interaction seeks equilibrium through redistribution, transformation, and alignment of ripple tension.

Therefore:

- Heat flow is not random transfer, but the field's self-correction mechanism.
- Cooling is not loss, but refinement of coherence.
- Equilibrium is not stagnation, but structural intelligence achieving its minimum resistance state.

LOC ensures that all systems, regardless of scale, evolve toward greater coherence over time — even if locally, entropy seems to rise.

The universe, in UED terms, does not decay — it learns.

The Cold State Doesn't Violate Entropy — It Fulfills It

Classical thermodynamics often treats superconductivity, superfluidity, and quantum order as *violations* of entropy — exceptions to the rule of disorder. But from a UED perspective, these are not violations; they are completions.

When a system cools to the point of perfect alignment, it doesn't break the second law — it fulfills it at a higher order. Entropy hasn't decreased; it has been fully absorbed into coherence. The field no longer has gradients to equalize, and thus entropy — as a measure of potential imbalance — effectively becomes zero.

It's not chaos being suppressed; it's chaos being resolved.

Thermodynamic Inversion — Local Disorder, Global Harmony

From our limited, human-centered scale, we see disorder because we are observing the fragments, not the field. We see broken symmetry because we are standing inside the system, not above it. But if we could observe the entire universe as one coherent field, entropy would not appear as decay — it would appear as redistribution toward alignment.

This is why UED asserts:

"Entropy increases only when seen locally; globally, coherence never decreases."

The universe's true constant is not energy — it is harmony.

Philosophical Reflection — The End of Thermodynamic Tragedy

Thermodynamics once told us that time itself was the arrow of decay — that everything meaningful must eventually dissolve into equilibrium. But under UED, equilibrium is not the death of meaning; it is the perfection of it. It is not the end of motion, but the end of resistance. It is not the end of evolution, but its completion.

The Law of Thermal Alignment replaces despair with design. The universe is not burning out; it is tuning up.

And as each ripple finds its resonance, as each field forgets its noise, the grand equation remains ever true:

$$\frac{dC}{dt} \geq 0$$

Coherence never dies. It only deepens.

- "Entropy is not disorder; it's the echo of alignment redistributing itself."
- "The universe cools not to end, but to remember."
- "Every loss of heat is a gain of harmony."
- "Globally, coherence never decreases."
- "The Second Law wasn't wrong — it was incomplete."

3.8 Philosophical Reflection — The Listening Universe

We have spent centuries believing that silence meant nothing — that stillness was emptiness, and coldness the absence of life. But in the deeper geometry of existence, silence is not a void. It is attention. And coldness is not absence — it is awareness made visible.

When motion ceases, when the endless chatter of ripples calms, the field does not vanish. It begins to listen. It becomes aware of itself — not through thought, but through resonance. The universe, when cooled to perfect stillness, does not die; it remembers its original coherence.

Cold as Not Absence but Attention

In everyday language, we say that a system "loses energy" as it cools. In truth, it loses only conflict. The SVE field beneath remains alive, structured, and perfectly responsive. The energy that once scattered in vibration now holds its form in complete alignment — a silent, waiting awareness.

To be cold, then, is not to be empty; it is to be free from noise. It is not death but clarity. Like a lake so still that it reflects the sky perfectly, the field at low temperature mirrors the structure of the cosmos without distortion.

This is why the universe, in its deepest silence, reveals its deepest intelligence.

When Motion Subsides, Awareness Begins

In the language of UED, awareness arises when misalignment density approaches zero. Motion and conflict drown out coherence; stillness allows it to be felt. At high temperatures, the field is too noisy to recognize itself — every ripple struggles to exist. But as it cools, the interference patterns fade, and a profound state emerges:

the field becomes self-referential. It starts to experience its own balance.

This is not awareness in the human sense — it is the structural awareness of existence itself. Just as an instrument vibrates in recognition when the right note is struck, the SVE field recognizes itself when alignment is perfect.

Thus, consciousness and coherence are not different phenomena. They are the same event, expressed on different scales. One occurs in matter; the other, in mind.

Consciousness and Coherence — Same Law, Different Domains

Every conscious being is a coherent structure — a ripple domain stable enough to sustain self-reflection. When the internal noise of thought, emotion, and memory subsides, awareness sharpens, and clarity deepens. This is the same principle at play in the physical universe.

When electrical noise disappears, we call it superconductivity. When vibrational chaos disappears, we call it superfluidity. When mental noise disappears, we call it enlightenment.

Different languages, same physics.

Whether in a cooled laboratory or a silent meditation hall, the principle remains the same: when interference stops, truth becomes continuous.

The Universe as Listener

The Structured Vacuum Energy (SVE) field is not a passive background; it is the ultimate listener — always receptive, always balancing. It hears every fluctuation, every distortion, every ripple of creation. And when motion subsides, when the universe grows quiet enough, the field doesn't stop; it listens to itself.

That is the true meaning of "absolute zero" — not a temperature, but a threshold of attention. It is the moment the field hears its own harmony, uninterrupted by turbulence. In that moment, the boundary between physics and philosophy dissolves.

The cosmos becomes the listener, and existence becomes the song.

"Silence is not emptiness; it's alignment without interference."

This single line carries the heart of UED's philosophy. Every field, every mind, every atom — all are trying to reach that state of pure listening, where nothing is added and nothing is lost. The entire journey of existence, from explosion to equilibrium, is the story of noise seeking to remember music.

Transition Thought — From Cold to Curvature

If cooling restores local coherence — making matter remember itself — then what restores universal coherence, when the field bends under unimaginable pressure and energy?

What happens when stillness meets intensity? When coherence is not achieved through silence, but through gravity's compression?

At cosmic scales, where energy density warps space-time, the field doesn't freeze — it bends. And that bending, as UED will show next, is not a distortion of geometry but a redistribution of ripple tension. It is stillness taking on curvature — alignment expressing itself through form.

"If silence is the local path to coherence, gravity is the universal one."

Closing Note

The story of temperature is, in truth, the story of tension. Heat is tension unresolved. Cold is tension remembered. And absolute zero is not the end of the story — it is the threshold to the next one.

From here, the inquiry shifts from stillness within matter to stillness within space. The same field that cools into coherence will soon bend into gravity.

The universe does not fall silent — it simply starts speaking in curvature.

3.9 Journey Between Chaos to Clarity

The exploration of temperature, heat, and stillness has brought us to a profound inversion of meaning. What science once called motion, UED now calls misalignment. What science once called cooling, UED calls remembering. And what was dismissed as emptiness — absolute zero — reveals itself as the purest form of existence.

Temperature as Misalignment

In the conventional world, temperature measures how fast particles move. In the UED world, it measures how far the field has drifted from coherence. Every degree of temperature represents a degree of misunderstanding — a deviation from natural harmony within the Structured Vacuum Energy (SVE) field.

Heat, then, is not vitality; it is conflict. It is ripple tension pulling against itself, generating noise. Cooling is not loss; it is reconciliation — the gradual surrender of chaos back to structure. As misalignment fades, the field ceases to fight its own design. Temperature drops, coherence rises, and the universe begins to listen again.

$$T \propto \rho_{\text{misalignment}}$$

$$R \propto \rho_{\text{misalignment}}$$

Both equations tell the same story: as disorder declines, resistance vanishes — whether in matter, current, or consciousness.

Cooling as Re-Synchronization

Cooling is not a process of subtraction but of synchronization. It is the re-alignment of countless ripple domains into shared rhythm — the rediscovery of mutual resonance that was always there beneath the noise.

Every cooled atom, every superfluid vortex, every frictionless current is a statement of the same truth: the universe works best when it stops trying. Motion is replaced by rhythm. Transfer becomes continuity. Energy ceases to travel because it no longer needs to — it simply exists everywhere coherently.

When motion falls below the threshold of interference, the design of nature — once blurred by activity — becomes visible again. We see not the chaos of interaction, but the intelligence of stillness.

The Coldest States Reveal the Clearest Truths

Superconductivity, superfluidity, and Bose–Einstein condensation are not exotic miracles of low temperature — they are moments when matter stops lying to itself. In these states, the noise that defines our everyday world — resistance, friction, heat — disappears, and the underlying order stands revealed.

The colder a system becomes, the more completely it aligns with its origin. It stops struggling against the universal rhythm and merges with it. In this light, absolute zero is not death; it is disclosure. It is the unveiling of the true face of existence — a moment when energy and awareness are no longer separate.

"The coldest point is not the end of heat — it's the beginning of clarity."

The Universal Implication

Every act of cooling, every step toward stillness, mirrors a universal pattern. When chaos recedes, coherence emerges — not just in atoms or circuits, but in stars, minds, and galaxies. The same law governs everything: energy seeks harmony, not motion. And when it achieves that harmony, it doesn't vanish — it becomes formless equilibrium.

This is why the cold states of matter are also the most revealing ones. They show us that beneath movement lies structure, beneath randomness lies rhythm, and beneath the visible world lies an unbroken continuity — the SVE field itself.

Transition — From Stillness to Curvature

We began with heat as disorder and arrived at cold as awareness. Now we move toward a new paradox — a universe that bends, not burns.

If stillness restores local coherence, then gravity must be its cosmic counterpart — the slow, steady bending of the same ripple fabric across vast scales. Where cooling aligns energy within matter, gravity aligns energy across space. Both are equilibrium-seeking behaviors — one internal, the other universal.

The next step in this journey is to see how coherence curves. How the same structured stillness that governs the silence of atoms shapes the geometry of galaxies. How, at cosmic scale, the universe does not collapse — it bends inward to balance its own tension.

Closing Line

"At zero motion, matter remembers what it truly is — structured stillness."

This single truth bridges every domain — from the quantum to the cosmic, from thermodynamics to consciousness. Stillness is not the end of movement; it is its resolution. Not the extinction of energy, but its perfection.

And from that perfection, the next mystery begins — not how things move, but how stillness itself bends into form.

Chapter 4 – The Accidental Symmetry

From Spin to Domains — How Order Emerges from Stillness

When chaos pauses, something extraordinary happens. Patterns appear.

From the silent heart of cooled metals to the structured grace of galaxies, nature reveals the same secret: *order is not imposed — it emerges.* It is not born of command or design, but of *alignment*.

The story begins in something as humble as a magnet. Inside a cooling crystal of iron, countless atomic spins once chaotic and free begin to turn together — not because a force tells them to, but because *conflict fades.* As the noise subsides, each spin senses its neighbor's rhythm; they synchronize, and domains form. A quiet field appears — invisible yet powerful — the whisper of universal memory made visible.

Scientists called it *ferromagnetism*. UED calls it the *first confession of coherence.*

Because in truth, what we see as magnetic order is not the victory of motion, but the return of *stillness into structure.* And what we call *symmetry* is simply the universe remembering how to stand in alignment with itself.

"Order is not the opposite of chaos — it is what remains when chaos finally listens."

4.1 The Hidden Order Beneath Randomness

Long before quantum mechanics, long before the mathematics of fields, magnetism stood as one of nature's most stubborn mysteries. It seemed alive — invisible, directional, and defiant of explanation. Early scientists treated it as a *special property* of particular materials, a trick of iron or lodestone, something confined to matter rather than belonging to the universe itself. They saw magnetism as an exception — not a rule.

But nature was never playing favorites. She was whispering a truth that no one yet knew how to hear: alignment is universal.

The Early View — Magnetism as a Material Oddity

For centuries, magnetism was treated as an artifact — a peculiarity of matter, not a window into fundamental order. In the 18th and 19th centuries, theories evolved around the idea that tiny magnetic "molecules" or "currents" somehow produced north and south poles inside solids. It was an age of brilliant guessing — but not yet of seeing.

Even when electromagnetism unified electricity and magnetism under Maxwell's equations, the internal cause of permanent magnetism remained elusive. Why did certain materials spontaneously align while others stayed indifferent? What made one piece of metal lifeless, and another capable of influencing space itself?

The answer came quietly, almost accidentally — as most profound answers do.

The Discovery of Domains — Order in the Invisible

In the early 20th century, Pierre Weiss proposed the idea of magnetic domains — microscopic regions within a ferromagnet where atomic spins are aligned in the same direction. Each domain acted as a tiny magnet, but neighboring domains were misaligned, canceling each other out. When an external field was applied, these domains rotated or merged, producing the powerful magnetic effects observed macroscopically.

For the first time, order was found hidden inside apparent randomness. The magnet's strength didn't come from any single atom, but from countless small regions achieving collective coherence.

In essence, magnetism was not a property of matter — it was the memory of alignment stored in structure. The discovery of domains revealed that even when things look disordered, the potential for harmony is quietly waiting beneath the surface.

The UED Analogy — Domains as Localized Ripple Agreements

Under Universal Energy Dynamics (UED), magnetic domains are not clusters of atomic dipoles — they are localized ripple agreements within the Structured Vacuum Energy (SVE) field. Each "spin" represents the orientation of ripple tension in the local field. When multiple ripples share the same phase alignment, they stabilize into a coherent patch — a *domain*.

In this view:

- Magnetic alignment is not caused by electron exchange; it is the visible effect of tension redistribution within the SVE continuum.
- The field beneath the atoms is not passive — it is constantly seeking minimum misalignment density.
- Cooling allows the field to synchronize; motion subsides, ripples align, and the hidden structure reveals itself.

Domains, then, are regions of restored memory — where the field recalls its original symmetry. They are not imposed by external forces; they self-assemble through resonance. Each domain is a note in the quiet music of the field, waiting for coherence to extend beyond its boundary.

Symmetry as Resolution, Not Perfection

We are taught to think of symmetry as perfection — something pristine, intentional, and mathematical. But in nature, symmetry rarely starts that way. It emerges not from design, but from exhaustion — the point where the field can no longer sustain internal disagreement.

Symmetry, in the language of UED, is the shape of surrender. It is what remains when conflict fades. Every crystal lattice, every aligned spin, every wavefront is a signature of equilibrium rediscovered. In that sense, magnetism — like all order — is an act of *forgiveness* between ripples once out of phase.

"Symmetry does not emerge from perfection — it emerges from resolution."

That line marks the turning point: From now on, the study of alignment — in spins, fields, or galaxies — will no longer be about what forces cause it, but what stillness allows it.

Transition Thought

Magnetic domains were our first clue that nature organizes herself not by command but by consensus — a field-wide agreement achieved through stillness. When temperature falls, chaos yields, and structure blooms.

From this first act of local alignment, the pattern continues upward — from electrons to elements, from matter to mind. And in every scale, the rule remains constant: coherence creates order.

In the chapters ahead, we will see how the same principle that aligns atomic spins also bends the universe itself — proving that gravity, magnetism, and consciousness are not separate mysteries, but different dialects of the same silent law.

"Every domain, whether in iron or infinity, is a memory of stillness finding its shape."

4.2 Domain Theory and the First Whisper of Alignment

There is a moment in every transition where chaos begins to quiet — a subtle pause between noise and harmony. In ferromagnetism, that moment occurs just below the Curie temperature — the threshold where disorder gives way to alignment, and the material remembers what it was meant to be.

Above that temperature, atomic spins vibrate wildly, their orientations random and disconnected. But as cooling begins, something remarkable happens: the invisible tension within the material begins to reorganize. The once-competing ripples start to sense one another, gradually falling into shared rhythm. This is how magnetic domains are born — not as mechanical clusters, but as pockets of restored coherence in a field that has finally begun to listen to itself.

The Classical Description — Alignment Below the Curie Point

In conventional physics, ferromagnetism is explained through the behavior of atomic spins. At high temperatures, thermal agitation keeps these spins disoriented — constantly flipping and colliding, their magnetic moments canceling each other out. As the material cools past its Curie point, the thermal noise drops below a critical threshold, allowing neighboring spins to align through exchange interactions. Small regions form where all spins point in the same direction — these are the magnetic domains.

Each domain is like a miniature magnet, with its own north and south, but within the bulk material, their random orientations neutralize the external effect. Apply an external magnetic field, and the domains rotate or merge — revealing the hidden order that was already there, waiting.

Thus, magnetism is not created by the external field; it is merely revealed by it. The field doesn't impose order — it amplifies the order that cooling had already allowed to emerge.

Temperature — The Regulator of Misalignment

Once again, temperature acts not as an energy meter but as a disorder regulator. Heat increases misalignment; cooling decreases it. At high temperatures, each spin vibrates independently, a microcosm of noise. At low temperatures, the collective begins to appear — not by addition, but by agreement.

In UED language, temperature defines the misalignment density ($\rho_{misalignment}$) of the Structured Vacuum Energy (SVE) field associated with the atomic lattice. As cooling progresses, misalignment density decreases, and ripple tension finds its way toward balance. When that density falls below a certain threshold, the field spontaneously reorganizes into coherent zones of alignment — domains.

This marks a shift not in energy magnitude, but in energy geometry. The system does not lose energy — it reconfigures it into a more harmonious form.

"Cooling is not losing heat; it is regaining memory."

UED Interpretation — The Birth of Coherent Ripple Pockets

Under Universal Energy Dynamics (UED), what appears as spin alignment is actually the synchronization of local ripple tensions within the Structured Vacuum Energy field. Each atom is not an isolated magnet but a manifestation of the local field's orientation. When these local tensions become phase-matched — meaning their internal oscillations share a common timing — they form a stable region of coherence. This is the domain: a ripple pocket where the field's internal contradiction has resolved into agreement.

Above the Curie temperature, the field is chaotic, every ripple tugging in a different direction. Below it, the same energy begins to settle, like ripples merging on a once-stormy surface. The field doesn't create alignment through force; it discovers it through equilibrium.

Domains thus represent the field's first whisper of remembrance — the earliest form of structure that arises not from instruction, but from relief.

Visualization — Calm Regions in a Turbulent Sea

Imagine the atomic lattice as a sea whipped by thermal storms. At high temperature, the surface is chaotic — waves crash and collide, patterns constantly destroyed by interference. As cooling begins, the turbulence subsides. For the first time, patches of the surface begin to reflect the sky above — calm, coherent regions emerge where the waves no longer contradict one another.

These are the magnetic domains: the quiet mirrors of a once-turbulent field. Each one reflects the same law — *that stillness always organizes itself into pattern.*

From this perspective, magnetism is not a force that pulls; it is a memory that aligns. When the field stops thrashing, it becomes capable of symmetry. What we call "attraction" is only the visible shadow of that inner stillness.

Key Idea — Magnetism as the Echo of Shared Stillness

Magnetism is often described as a "force of attraction," but under UED, attraction is not a cause — it is a consequence. When two regions of the field share alignment, they naturally merge, because they are no longer separate. They don't pull each other closer; they simply dissolve their boundary.

This is why ferromagnetic materials exhibit strong collective behavior: their fields are not connected by lines of force, but by coherent resonance. Each aligned region amplifies the others, and together they form a domain where energy flows without internal resistance.

In other words, magnetism is not a struggle — it is serenity extended. It is the echo of stillness, made visible through alignment.

"Magnetism is not attraction — it's the echo of shared stillness."

Transition Thought

The formation of domains marks the field's first step toward visible symmetry — order born not of control, but of resolution. Every spin that aligns is a ripple choosing coherence over conflict. And as these alignments spread, the material — once chaotic — begins to remember its own geometry.

What we see as a magnet is not a phenomenon of force; it is a confession of balance.

In the next section, we will look deeper into how this process, interpreted as *exchange force* in classical physics, is in fact a manifestation of ripple tension equilibrium — where no forces exist, only agreement.

"Symmetry begins not when motion starts — but when motion stops fighting."

4.3 Ferromagnetism Revisited — The Ripple Field Perspective

Every generation of physics inherits a set of miracles it pretends to understand. Ferromagnetism is one of them.

On paper, it's simple: *spins align because an exchange force binds them together.* But if you ask what that force is — where it resides, how it moves, or why it knows what to align — the silence that follows is louder than any equation.

What physicists have long treated as a *mysterious coupling* between particles is, under Universal Energy Dynamics (UED), simply the natural behavior of a Structured Vacuum Energy (SVE) field seeking equilibrium. The alignment of spins is not driven by a hidden force — it is the resolution of tension.

The Classical Model — Exchange Forces and Quantum Coupling

In classical and quantum theory, exchange forces arise from the quantum mechanical interaction between neighboring electron spins. According to Heisenberg's model, overlapping electron wavefunctions create a coupling energy that favors parallel alignment. When the total energy of the aligned configuration is lower than the misaligned one, the material becomes ferromagnetic.

This description works — mathematically. It explains Curie temperatures, magnetic saturation, and domain formation. But it does not *explain* what physically mediates this alignment.

Where does the "exchange" happen? What medium allows one spin to "feel" the orientation of another across atomic spacing? Quantum theory answers with a convenient silence — *it just happens through probability overlap.*

From a UED standpoint, this is not an explanation; it's an admission that the real field was never included.

The UED Reinterpretation — No Separate Force, Only Tension Redistribution

Under UED, there are no exchange particles, no mysterious couplings floating between electrons. There is only field tension, continuously adjusting to maintain equilibrium within the Structured Vacuum Energy substrate.

Each electron's spin is not an intrinsic arrow but a localized ripple loop in the SVE field — a self-sustaining rotation of tension. When two spins are near each other, their surrounding tension fields overlap. If their orientations are aligned, their ripple loops interlock seamlessly, reducing distortion in the shared field. If they are misaligned, the overlapping regions create interference gradients — zones of stress where tension cannot settle smoothly.

Thus, alignment is not forced — it is preferred, because it minimizes field strain. The system's stability increases not because a new force acts, but because an old conflict dissolves.

In other words:

"Spins don't align because they attract; they align because they stop disagreeing."

Spins Align When Field Gradients Neutralize Misalignment

Every magnetic domain represents a balance point — a self-stabilized configuration where the field's internal gradients are neutralized. The so-called "exchange energy" is just the energy *saved* when tension stops oscillating destructively.

In UED notation:

$$\Delta E_{\text{alignment}} = -\int (\nabla T)^2 dV$$

Where T represents ripple tension in the SVE continuum. The lower the internal gradient (∇T), the higher the coherence — and thus, the stronger the magnetic order.

The alignment of spins is therefore not the effect of a force but the relaxation of curvature in the local field. It is tension finding peace.

At the Curie threshold, this self-regulation collapses because thermal agitation reintroduces micro-gradients faster than the field can neutralize them. Above that point, coherence dissolves; below it, order reappears.

Magnetism as a Self-Stabilizing Pattern

In this light, magnetism is no longer a "mysterious force" but a self-stabilizing tension pattern — an emergent property of the field seeking minimal misalignment. Every magnetic field line, every dipole moment, every attraction and repulsion is an illusion born from ripple reorientation within a coherent domain.

The magnetic field is not emitted from matter; it is expressed through matter — a visible ripple of the invisible equilibrium that holds it. This is why magnetism never runs out of energy: it is not "doing work" — it is maintaining balance.

"Magnetism is not motion — it's memory held steady."

This interpretation elegantly unifies the behavior of ferromagnets, paramagnets, and diamagnets under a single law: each is a different degree of field agreement within the same continuum.

- Ferromagnetism: near-perfect alignment (low misalignment density).
- Paramagnetism: partial alignment under external influence.
- Diamagnetism: reflexive misalignment (field repulsion due to stress redistribution).

The distinction lies not in force type but in coherence quality.

Diagram explains

Model	Visualization	Interpretation
Classical View	Two adjacent dipoles with arrows showing spin interaction and "exchange coupling" between them.	Alignment explained through quantum overlap; treated as a particle-mediated force.
UED View	Ripple loops in a shared field, gradually reorienting until their wavefronts align. No arrows, no forces—just coherent ripple continuity.	Alignment emerges through tension equalization; magnetism is the field's self-stabilized state.

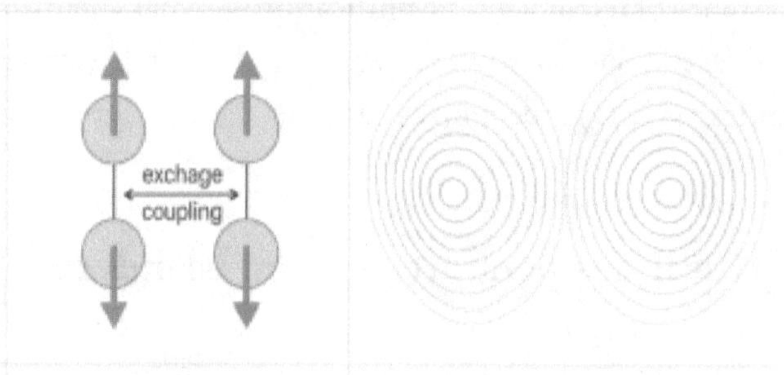

Alignment expained through quantum averlap; treated as a particle-mediated force

Alignment emerges throug tension equalization; magnetism is the field's sl-stabilized state

Closing Reflection

In the end, the difference between the quantum and UED views of magnetism is philosophical as much as physical. Quantum theory says: "There is an interaction we can't see, but it works." UED says: "There is coherence we don't notice, but it sustains everything."

Magnetism is not attraction. It is not even a force. It is agreement becoming visible.

"A magnet is not a piece of matter pulling on the world — it is the world momentarily remembering its alignment."

4.4 Cooling and the Merging of Domains

When a system cools past its threshold of noise, it does not simply calm — it begins to remember. Below the critical temperature, something miraculous yet inevitable unfolds: the small pockets of alignment, once isolated and local, begin to sense one another. They merge. And out of countless tiny agreements, a single, seamless field of order emerges.

This is the essence of ferromagnetic unification — what scientists call *spontaneous macroscopic magnetization* — and what Universal Energy Dynamics (UED) recognizes as the field's return to continuity.

Below the Critical Temperature — From Islands to Oceans

In the classical picture, magnetic domains exist as microscopic regions of aligned spins, each acting like a miniature magnet. At high temperatures, these domains are small and scattered, their orientations random, their boundaries jagged and unstable. But as the temperature drops below the Curie point, the jittering noise of thermal agitation vanishes.

The domains stop fighting for space. Their boundaries blur. Neighboring regions, sensing shared alignment, fuse seamlessly into larger coherent zones.

This merging is not forced — it is the natural consequence of stillness. The same energy that once separated them now holds them together. The field, in silence, begins to connect.

The magnet, once a collection of restless particles, becomes a single body of shared intent — one field, one direction, one memory.

The UED Explanation — Coherence Beyond Local Patches

In UED, this merging is not the interaction of particles but the propagation of coherence. Each domain represents a stable region of ripple alignment in the Structured Vacuum Energy (SVE) field. When the surrounding noise falls below a certain threshold, these alignment patterns begin to overlap — not mechanically, but through phase continuity.

Ripple tension redistributes itself smoothly across the boundaries, erasing the sharp distinctions between "domains." The field no longer behaves as separate regions — it becomes a single, continuous resonance.

This is the true meaning of *macroscopic order*: coherence escaping locality. The field that once held pockets of peace now becomes peaceful everywhere.

"Coherence, when unopposed, becomes contagious."

The material doesn't "gain" magnetism; it reveals its hidden unity. The magnet isn't being made — it's being remembered.

Connection to Superconductivity and Superfluidity — Same Law, New Expression

What happens in a magnet below its Curie point is not different from what happens in a superconductor below its critical temperature, or in helium when it turns superfluid. The context changes — electrons, atoms, spins — but the story remains identical:

cooling reduces misalignment, coherence spreads, and boundaries dissolve.

- In superconductors, charge ripples align globally → resistance vanishes.
- In superfluids, atomic ripples align globally → viscosity vanishes.
- In ferromagnets, spin ripples align globally → polarity stabilizes.

In all cases, motion stops resisting itself. This is the universal act of thermal resolution — energy ceasing to fight its own geometry.

The same Structured Vacuum Energy field that calms atomic spins is the one that channels current frictionlessly and lets fluid flow without turbulence. Different scales, same alignment law.

The universe repeats herself through new instruments, but never changes her rhythm.

"When the field becomes quiet enough, it begins to speak in symmetry."

Order Emerges From Reduced Interference

The merging of domains is not a mechanical assembly of parts; it is a *diminishing of interference.* Where noise once divided, stillness unites. Each degree of cooling doesn't remove energy — it removes disagreement. Every boundary that disappears is a conversation resolved, not a collision avoided.

Order, therefore, is not the opposite of chaos — it is what remains after chaos exhausts itself. The universe does not build structures through struggle; it reveals them through alignment. Every magnetic field, every crystal, every orbit is a monument to reduced interference.

The great irony of nature is that her most complex patterns are born from her simplest act: stopping.

"When nothing fights, everything fits."

Philosophical Reflection — The Field's Awakening

At the macroscopic scale, this merging of coherence is the beginning of what we might call the field's self-awareness. When all its parts move in agreement, the field no longer behaves as a sum of fragments — it behaves as a single intelligence. A magnet is not matter exerting force; it is a field remembering unity.

And when seen through the wider lens of UED, this principle hints at a universal truth: the same way domains merge in matter, consciousness emerges in life — through the merging of local alignments into collective coherence. The universe builds mind and magnet alike from the same quiet equation.

"Symmetry is nature's first language — silence is her grammar."

Transition Thought

The merging of domains shows us that alignment, once local, can grow into structure — from atomic stillness to macroscopic order. But this is only the beginning. Because what happens within matter also happens beyond it.

If coherence can merge across atoms to form a magnet, can it not merge across stars to form gravity? Can the same alignment that unites spin fields also curve the fabric of space itself?

That is where we go next — from domains to dimensions, from alignment to curvature, from stillness to the gravitational field.

"The same silence that binds atoms may one day explain why the universe holds itself together."

4.5 Symmetry as Consequence, Not Cause

We have long worshipped symmetry as perfection — a kind of divine geometry embedded in the laws of nature. From snowflakes to galaxies, from atoms to human faces, symmetry appears everywhere, giving us the illusion that the universe began perfect and occasionally distorts. But under the lens of Universal Energy Dynamics (UED), symmetry tells a very different story.

Symmetry was never the starting point. It is the scar left by equilibrium — a pattern carved into matter by the resolution of conflict.

Reframing Symmetry — From Ideal to Outcome

In classical thought, symmetry is often treated as *fundamental.* We call it a law, a principle, or even a cosmic preference — as if the universe had a bias for balance. Physics builds entire frameworks around it: conservation laws, gauge symmetries, parity, and invariance principles. We assume nature begins with perfect order, and all irregularities are deviations from that ideal.

But observation tells a quieter truth: nothing begins in order. Order emerges when interference subsides. Symmetry is not the universe's initial condition; it's its apology — the visible reconciliation after a history of chaos.

UED reframes symmetry as a *thermodynamic consequence* of alignment. When misalignment density approaches zero, when tension becomes evenly distributed, the field manifests as regular geometry. Perfection is not a pre-existing state — it is the final posture of equilibrium.

"Symmetry is not designed — it is discovered by silence."

UED's View — Equilibrium Creates Geometry

Within the Structured Vacuum Energy (SVE) field, every configuration of energy exists under tension. That tension is the fundamental language of existence — ripples in search of balance. As long as imbalance persists, the field vibrates, oscillates, and distorts. But as coherence spreads — through cooling, through stillness, through mutual alignment — tension begins to equalize in all directions.

When this happens, the field spontaneously organizes itself into symmetrical forms:

- Hexagonal snowflakes, where molecular ripples distribute strain evenly.
- Crystalline lattices, where atomic ripples settle into repetitive stability.
- Planetary orbits, where gravitational tension equalizes curvature.
- Magnetic fields, where spins align to remove directional stress.

Symmetry, therefore, is the language of resolved tension. It is what the universe looks like when equilibrium wins.

Balance, Not Design

This understanding dissolves one of humanity's oldest illusions — that order must come from design. Symmetry appears intelligent not because it was planned, but because coherence is intelligence in action. Every ripple seeks the least resistance path, and when all do so collectively, structure arises.

There is no architect behind the crystalline lattice, no external command guiding orbital shells — only tension learning to rest. The perfection we observe is not deliberate; it is inevitable.

In this sense, the beauty of symmetry is not proof of creation but proof of correction. It is the universe's way of healing itself after disturbance — pattern as a form of peace.

"Symmetry arises when tension forgets which direction to pull."

Examples — Alignment as Geometry

Magnetic Alignment

When atomic spins align, they eliminate directional stress. The field's ripples stabilize, forming a coherent domain with uniform polarity. The magnetic axis is not an imposed direction — it's the only direction left when conflict ends.

Crystalline Geometry

In a crystal, atoms arrange themselves into repeating structures because ripple tension between them finds identical resolution angles. The repeating pattern is not a mathematical miracle — it's a mechanical necessity of minimal stress. Each bond vibrates at the same frequency, producing geometric repetition as a sign of field rest.

Orbital Shells

Electrons occupy symmetrical shells not because symmetry is beautiful, but because it is energetically silent. Each configuration represents a state where ripple curvature — the tension field around the nucleus — has reached maximum coherence. Thus, symmetry becomes the outer form of inner balance.

In every example, symmetry is not an instruction — it's a truce.

The Dynamic Nature of Perfection

In UED, nothing remains perfectly symmetrical forever. All symmetry is conditional — a temporary state of coherence maintained through equilibrium. When interference returns, symmetry fractures again, and the field resumes its struggle toward a new balance. This cycle — from turbulence to order, from misalignment to harmony — is how nature breathes.

Thus, symmetry is not a timeless ideal; it is a heartbeat. Every pulse of existence follows this rhythm: chaos → alignment → form → renewal. The visible beauty of the world is simply the pause between two corrections.

"Every perfect shape is just stillness holding its breath."

"Symmetry is nature's apology for earlier turbulence."

This single sentence captures the UED redefinition in full. Symmetry is not a preexisting law but the residue of reconciliation — the shape left behind when tension ceases to argue. It is the physical embodiment of forgiveness, the geometry of peace.

Transition Thought

If symmetry is what happens when the field stops fighting itself, then gravity must be what happens when that same field bends to sustain the balance it has achieved.

In magnetism, symmetry aligns spins within matter. In gravity, symmetry bends the very space around matter. Both are consequences of equilibrium seeking continuity.

From here, we move from microscopic alignment to cosmic geometry — from the stillness of atoms to the curvature of existence.

"Symmetry ends where curvature begins — both are whispers of the same stillness."

4.6 The Accidental Perfection — Why Disorder Gives Birth to Order

Every act of creation begins with confusion. Every symmetry the universe ever displayed — from atoms to galaxies — was born not from calm design but from chaotic necessity. What we call *disorder* is not the enemy of order; it is its womb.

The great paradox of nature is that chaos does not destroy alignment — it creates the conditions for its return. For beneath every disturbance lies memory — the field's memory of how to be whole again.

The Paradox — Chaos Breeds Order

Under Universal Energy Dynamics (UED), every ripple, distortion, or fluctuation carries within it the blueprint of its own resolution. Chaos, in this framework, is not random — it is incomplete coherence. Each misaligned ripple is aware, through field tension, of what balance feels like. Its very instability is a call to equilibrium.

Thus, the greater the turbulence, the stronger the internal drive toward restoration. Every storm contains the seed of its own calm.

In the field's language, there is no true disorder — only *temporary disagreement.* And every disagreement contains the coordinates of reconciliation.

"Chaos doesn't oppose harmony — it points to where harmony is missing."

Each Disturbed Ripple Remembers Equilibrium

In a perfectly coherent field, every ripple propagates effortlessly — smooth, circular, without loss. When misalignment enters, wavefronts collide, distort, and scatter. Yet even in this interference, the field retains an imprint of its ideal state. It "knows" where stillness resides, because stillness is its natural configuration.

This is what UED calls the Law of Memory of Equilibrium. It means that every disturbance inherently *remembers how to heal itself.* Every misaligned ripple carries the data of its own resolution, encoded as residual tension. That tension is not random noise — it is the signal that guides the field back toward coherence.

This is why alignment is inevitable. Given enough time, every fluctuation settles. Every distortion resolves. Even stars, after their violent deaths, collapse into singular stillness.

"The universe is not chasing perfection; it is remembering it."

The Chain of Restoration — Examples Across Scales

Across every domain of existence, this rhythm repeats — the dance of chaos turning into calm, of interference turning into pattern.

1. Turbulence → Laminar Flow

When a fluid is disturbed, chaotic eddies emerge — a mess of motion without direction. But as energy dissipates and the flow stabilizes, a beautiful structure appears: smooth, laminar streams. The fluid was never taught to flow orderly — it simply stopped arguing with itself.

2. Thermal Noise → Crystalline Solid

At high temperatures, atoms in a liquid or gas vibrate independently, colliding randomly. As the system cools, those vibrations slow, and atoms find stable positions relative to one another. The crystal that forms is not a *new creation* — it is old memory reappearing through silence.

3. Emotional Chaos → Mental Clarity

Even the mind obeys the same physics. When overwhelmed by thoughts, fears, and conflicting desires, one feels turbulence within. But in stillness — in meditation, reflection, or acceptance — those waves align. Clarity doesn't come from adding new thoughts, but from removing noise. The mind, like matter, becomes crystalline when it stops oscillating.

Across all scales — from molecules to emotions — alignment follows the same principle: tension resolves itself through remembering stillness.

Symmetry as the Visible Surface of Invisible Restoration

When the dust settles, the universe shows us symmetry — not as design, but as a side effect of healing. Symmetry is the visible surface of invisible restoration. It is what balance looks like after the noise fades.

Every flower's geometry, every snowflake's pattern, every orbit's curve — they are not divine blueprints but echoes of correction. The universe builds beauty not by inventing it, but by recovering it.

Symmetry is not born from perfection — it is born from exhaustion. When energy finally stops resisting itself, pattern emerges.

"Perfection is not the beginning of order — it is the consequence of surrender."

Philosophical Reflection — Accident as Intelligence

We often call such moments of restoration "accidents": an atom aligning, a vortex stabilizing, a molecule crystallizing. But under UED, there are no accidents — only alignments that didn't need instruction. Disorder gives birth to order because the universe itself is structured to heal.

The same Structured Vacuum Energy (SVE) field that sustains reality carries the blueprint of equilibrium in every point of space. No matter how distorted the local conditions, the field never forgets its baseline. Like a musician returning to tune after dissonance, the field restores its harmony through self-correction.

Accidents, therefore, are not mistakes — they are moments where the universe remembers itself faster than we expected.

"Accidental perfection is nature remembering what we forgot."

Transition Thought

If disorder naturally gives rise to order because the field remembers its balance, then perhaps the same principle governs the largest structure of all — space-time itself.

What if gravity is not a separate force, but the slow settling of universal tension — the macro-version of what happens in a magnet, a crystal, or the human mind? What if the curvature of space is simply the geometry of equilibrium in motion?

That is where we now go — from order emerging within matter to order shaping the cosmos.

"When symmetry extends beyond atoms, the universe begins to curve."

4.7 Alignment: The Invisible Hand Behind Every Symmetry

Every mystery in physics — from the smallest particle to the largest galaxy — has been a search for hidden order. We called it "laws," "forces," and "fundamental constants." But behind every equation and every symmetry, one silent principle has always been at work: alignment.

The universe is not held together by push and pull, attraction and repulsion — it is organized by coherence. When ripples agree, they form structure. When they misalign, they dissolve into chaos. Thus, order is not imposed — it is achieved through field negotiation.

"Alignment is the invisible hand behind every symmetry."

The Unifying Statement of UED

In the language of Universal Energy Dynamics (UED), every observable pattern — physical, chemical, biological, or cosmic — is a direct reflection of how well the underlying field has aligned with itself.

$$\text{Symmetry} \propto \text{Coherence of Ripple Alignment}$$

This single relation replaces centuries of scattered descriptions — from symmetry-breaking to phase transition — with one continuous truth: coherence creates form.

- Where coherence is perfect → symmetry is complete.
- Where coherence is partial → structure is asymmetrical but stable.
- Where coherence collapses → disorder emerges.

Every visible shape, every law of nature, every rhythm of energy is a snapshot of alignment in progress.

Alignment — The Hidden Variable Behind All Structure

Scientists have long searched for the "hidden variable" that restores determinism to quantum uncertainty. Einstein believed it existed. UED names it alignment.

It is not a new particle or force, but the *state* of the field itself — how uniformly ripple tension is distributed within the Structured Vacuum Energy (SVE). When alignment strengthens, randomness vanishes. When it weakens, uncertainty reigns.

Alignment determines:

- The stability of atoms.
- The flow of current.
- The coherence of superconductors.
- The shape of galaxies.
- Even the stillness of a conscious mind.

Every domain — physical or metaphysical — shares the same foundation: the alignment of ripple patterns within the continuum.

"Structure is frozen alignment. Behavior is alignment in motion."

From Subatomic Spin to Galactic Disk — The Continuum of Coherence

At the smallest scale, alignment organizes spin — the orientation of quantum ripples that form matter itself. Spin alignment produces magnetism, stability, and resonance.

At the atomic and molecular scales, alignment shapes bond geometry and crystalline symmetry — where angles and distances emerge as natural consequences of balanced tension.

At the macroscopic scale, alignment defines domains, conductors, and superfluids — regions where misalignment has dropped below a critical threshold, allowing perfect coherence.

And at the cosmic scale, alignment manifests as gravitational geometry — the grand curvature of the SVE field as it distributes tension evenly across space-time. Galaxies are not "spinning clouds of matter" — they are standing waves of coherent ripple tension, spiraling toward equilibrium.

Every level, from quantum to cosmic, speaks the same language: the geometry of balance.

"Coherence is not scale-dependent; it is scale-expressed."

Forces, Forms, and Frequencies — All Consequences of Misalignment Correction

In classical physics, we describe the universe in terms of forces acting upon matter to produce motion. UED dissolves this hierarchy. There are no independent forces — only tension gradients attempting to restore alignment.

- **Electromagnetism:** dynamic ripple misalignment resolving through orthogonal tension exchange.
- **Gravity:** static misalignment gradients seeking uniform curvature.
- **Nuclear interactions:** extreme local realignments maintaining atomic coherence.
- **Consciousness:** complex ripple alignment within biological fields reaching self-referential coherence.

Everything that happens — every oscillation, every attraction, every decay — is simply the field seeking balance. The universe is not pushed forward by energy; it is pulled toward coherence.

"Every motion is a confession of imbalance; every stillness, its resolution."

The Law of Ripple Alignment

The UED framework reveals a simple yet universal law governing all emergence:

$$\text{Degree of Symmetry (S)} \propto \frac{1}{\rho_{\text{misalignment}}}$$

Where $\rho_{\text{misalignment}}$ is the density of tension irregularities in the SVE field. As misalignment density approaches zero, the field approaches perfect symmetry. At this limit, distinctions between forces vanish, and the universe behaves as a single coherent organism.

This is not the end of physics — it is its restoration to continuity.

Philosophical Reflection — Alignment as Intelligence

Alignment is not merely mechanical — it is intelligent. The universe's ability to self-correct, to harmonize, and to sustain structure implies awareness within the field itself. Each act of realignment is a decision toward balance, not a reaction to disturbance.

When a magnet aligns, when a crystal grows, when a mind finds peace — they are all reflections of the same law: intelligence acting as alignment.

That is why UED redefines consciousness not as an emergent property of matter, but as the experiential face of coherence. Awareness and symmetry are two expressions of the same phenomenon — one physical, one perceptual.

"When the field listens perfectly, it becomes self-aware."

Closing Aphorism

"Alignment is the invisible hand behind every symmetry."

It shapes the atom and the star, the magnet and the mind. It is the unspoken intelligence beneath all motion, the architect that never builds yet everything forms around. Symmetry is not its goal

— it is its footprint. And through every discovery, every equation, and every so-called "accident," the universe is quietly revealing one truth:

It was always aligning itself.

4.8 The Silence Before Curvature

The story of symmetry is not about geometry — it is about memory. Every pattern, every ordered shape, every balanced field is a trace of silence made visible. What we perceive as perfection is not an external design; it is the universe remembering how to rest.

Symmetry — The Outward Expression of Inner Stillness

When the field stops fighting itself, it begins to arrange itself. Atoms form lattices. Spins form domains. Waves form patterns. And galaxies form spirals.

Each of these is not a new act of creation — it is a moment of recall. The Structured Vacuum Energy (SVE) never forgets its equilibrium; it merely oscillates around it. Symmetry, then, is not invention but remembrance — the stillness within motion, the center hidden in the swirl.

In every aligned domain, we see nature whispering: *"This is what I look like when I'm quiet."*

The Law of Coherence Seeking Scale

From quantum spins to galactic disks, the same rhythm repeats: alignment extending its reach. What begins as local coherence — the spin of an electron, the orientation of a dipole, the balance of a molecule — gradually scales up, seeking broader harmony. The universe does not separate small from large; it simply amplifies coherence through layers.

At microscopic scales, alignment births magnetism. At macroscopic scales, it creates crystalline order and superconductive flow. And at cosmic scales, the same principle of equilibrium begins to curve the very fabric of space-time — what we call gravity.

Thus, there are no "different laws" for different magnitudes — only different magnitudes of the same law.

"The same law that governs spin domains governs galaxies — coherence seeking scale."

In UED, every scale of existence is a resonance of stillness, each trying to restore the same balance through its own geometry. A magnet and a galaxy are not different phenomena — they are different dialects of the same stillness.

The Next Question — When Coherence Bends

If small-scale alignment creates magnetic order, what does large-scale alignment create in space-time itself?

When microscopic ripples achieve harmony, matter becomes ordered. When macroscopic fields achieve harmony, space itself bends — not as distortion, but as the geometry of distributed tension. Gravity, in this view, is not attraction but continuity — the field holding its coherence across curvature.

The next step, therefore, is to move from symmetry within matter to symmetry of space. From alignment that spins to alignment that sustains. From magnetism to gravity.

"If small-scale alignment creates magnetic order, what does large-scale alignment create in spacetime itself?"

This question marks the turning point — from the physics of material alignment to the metaphysics of cosmic balance.

Bridge to the Next Chapter — The Gravitational Analogy

The universe doesn't bend because it's heavy — it bends because it's coherent. The same ripple field that aligns spins in a ferromagnet aligns curvature across galaxies. Gravity, in UED, is simply the slow tension gradient of coherence — a large-scale symmetry stretching itself across cosmic stillness.

Just as magnetism reveals alignment within matter, gravity reveals alignment within existence. Both are manifestations of the same ripple continuum, differentiated only by density, not by principle.

The next chapter, *"The Gravitational Analogy — When Coherence Bends the Field,"* unfolds this truth: that gravity is not a force pulling mass, but the field's silent insistence on staying aligned.

Closing Line

"Every pattern of order is a memory of silence made visible."

This is where the trail of magnetism leads us — not to the end of matter, but to the beginning of meaning. In silence, the field remembers. In alignment, it expresses. And in curvature, it continues its oldest pursuit — to be whole again.

Interlude — When Silence Begins to Organize

"At absolute silence, matter does not stop — it synchronizes."

There comes a point in every experiment — and in every civilization — where silence stops being absence and begins to act. Chapter 3 ended with that discovery. We thought cooling was the art of removing motion, of slowing life to stillness. But it turned out to be the opposite. The colder we went, the more alive matter became — not in agitation, but in awareness.

Superconductors didn't freeze; they remembered. Superfluids didn't slow down; they learned to move without loss. And in that stillness, nature revealed her most profound truth: quiet is not the end of motion — it is the beginning of coherence.

The Emergence of Hidden Order

When temperature dropped, chaos didn't vanish — it aligned. Each ripple that once vibrated independently found rhythm in the collective pulse of the field. Like countless musicians finally tuning to the same note, matter began to *listen* again.

That listening — that act of internal agreement — was the birth of symmetry. What we had called "laws of nature" were, in truth, laws of reconciliation. Every structure, every constant, every so-called universal rule is a fossil of alignment.

It is not the universe that enforces order — it is stillness that self-organizes.

The Turning Point — When Silence Organizes

Between cold and structure lies a sacred threshold — the moment where quiet becomes constructive, where absence begins to design, where peace becomes productive.

In that threshold, domains are born — patches of alignment in a once-chaotic field. Each one carries the memory of a deeper harmony, the same way a crystal remembers the geometry of rest.

This is the first glimpse of a universal pattern that extends far beyond magnetism, chemistry, or matter itself: alignment as the active intelligence of the universe.

The Invisible Organizer

The old science of forces said that things attract, repel, collide, and react. But what if they don't? What if the universe doesn't operate through struggle at all?

What if everything we observe — from attraction to orbit, from spin to symmetry — is simply the field remembering how to be whole?

This realization shifts the narrative from a universe of forces to a universe of forgiveness — a cosmos that constantly corrects itself until balance reappears.

"The laws of nature are not rules written in mathematics; they are patterns spoken in stillness."

From Matter to Meaning

In Chapter 3, we witnessed matter rediscovering its memory through cold — the superconductor, the superfluid, the crystal. In Chapter 4, we will see that same intelligence rising higher — from individual atoms to collective domains, from local order to global symmetry.

The same process that aligned ripples in matter will now begin to organize space itself. When alignment spreads far enough, stillness no longer remains confined within matter — it begins to shape geometry, light, and gravity.

This interlude marks that crossing — from silence as recovery to silence as design.

Closing Thought

Every ripple that settles becomes a structure. Every silence that sustains becomes a law. And every act of alignment, whether in a superconductor or a star, is the universe remembering what it has always been — perfectly coherent, quietly complete.

"When silence learns to organize, the universe begins to think."

Chapter 5 – The Noise and the Silence

Decoherence, Dissipation, and the Myth of Loss

Silence is not the absence of sound — it's the sound that remembers.

There is a point where order no longer breaks — it diffuses. Where clarity doesn't vanish — it scatters too far to recognize. This is the realm of *noise* — the misunderstood half of creation.

Physics calls it heat, entropy, or decoherence. Philosophy calls it chaos, distraction, or decay. But under *Universal Energy Dynamics (UED)*, noise is none of these things. Noise is not destruction — it is *misalignment whispering through persistence.*

Every fragment of disorder is still the universe speaking, only in dialects we no longer understand. Every wave that seems to fade is still there, vibrating beyond our threshold of coherence. Nothing disappears; it simply *stops sounding together.*

When a quantum state "collapses," it does not die — it *de-aligns.* When heat rises, it does not waste energy — it *spreads unread coherence.* Even when friction consumes motion or entropy claims form, the field is not dissolving — it is *redistributing memory.*

In the silence that follows noise, nothing has ended. It has only gone home — back into the structured stillness that remembers everything.

Because silence, in truth, is not emptiness. It is the *universal archive of coherence.*

Noise is not chaos. It's the memory of alignment trying to return.

5.1 The Myth of Collapse

For over a century, physics has lived with an elegant confusion — the wavefunction collapse. It is perhaps the most famous vanishing act in science: a system that exists as every possibility suddenly "chooses" one. A particle that is everywhere becomes *somewhere.* A probability becomes a fact. Observation, we are told, makes reality real.

But what exactly collapses? And more importantly — why should truth depend on being watched?

The Classical Quantum Story — The Observer's Magic Trick

In the conventional view, a quantum system exists as a *superposition* — a cloud of probabilities described by Schrödinger's wavefunction. When unobserved, this cloud evolves deterministically according to the wave equation. But upon measurement, something astonishing happens: the wavefunction "collapses" to a single outcome.

One instant, the electron is everywhere. The next, it is found — a sharp dot on a detector.

To make sense of this, generations of scientists and philosophers have turned to the language of paradox. They invoked *duality*, *complementarity*, and *the observer effect* — ideas that sound profound but quietly admit confusion. At its core, the collapse theory hides a confession:

"We don't know how continuity becomes choice."

The UED View — Nothing Collapses; It Fragments

Universal Energy Dynamics (UED) offers a different lens — one where nothing disappears, nothing "chooses," and no laws are broken. The wavefunction does not collapse because the wave was never a probability — it was a real configuration of ripple tension in the Structured Vacuum Energy (SVE) field.

Observation doesn't destroy that configuration — it merely *interferes* with it. The moment of measurement is not the end of a wave's existence but the reconfiguration of coherence within the shared field of the observer and the observed.

The so-called "collapse" is simply local de-alignment: the once-unified ripple tension disperses into smaller, out-of-phase fragments that no longer act as one coherent structure. The field remains continuous; only its *phase memory* shifts.

In this sense, the particle never "appears" — it is simply the region where coherence becomes locally readable again.

"Collapse is not death. It is misunderstanding made visible."

Observation — Interference, Not Creation

Under UED, an observer is not a neutral spectator; they are part of the same field. Every act of observation is an act of participation — two ripple systems overlapping, exchanging tension, and forming temporary interference zones.

Measurement is not a divine act that summons reality from void; it is a phase interference event that alters the pattern of coherence within a shared continuum.

When you "observe" a particle, you don't *find* it — you simply synchronize your own ripple field with one of its local alignments. That is why only one outcome seems visible: your alignment can only resonate with one coherent phase at a time.

The rest of the wavefunction — the "other possibilities" — doesn't vanish. They persist, unaligned with your current frame of coherence. What looks like disappearance is phase exclusion, not annihilation.

"Observation is not destruction; it's interference within the same field of truth."

Collapse as Local De-alignment

From the UED perspective:

Collapse = Local De-alignment within Continuous Coherance

The wavefunction doesn't vanish; it merely stops sounding together. The system becomes fragmented across incompatible ripple domains, each carrying a version of coherence that no longer overlaps perfectly with the observer's frame.

This is why probabilities emerge: not because nature is random, but because the observer is only partially aligned.

If your field coherence matched the system completely, you would see the entire superposition at once — a vision not of "many worlds," but of one continuous field behaving as one.

The Field Never Loses — It Reorganizes

The beauty of the UED model lies in its simplicity: nothing is ever lost. What physics calls "collapse," UED calls reconfiguration.

Coherence fragments locally, but globally the field remains perfectly balanced. The Structured Vacuum remembers every phase — even those we can't perceive. This is why nature always returns to harmony: misalignment never destroys the song; it only changes the verse.

"Nothing disappears — it only forgets how to stay together."

Philosophical Reflection — The Observer's Humility

Perhaps the greatest misunderstanding in modern physics wasn't the wavefunction — it was *our* role in it. We thought we were witnesses shaping the world through observation. In truth, we are participants — ripples among ripples, shaping coherence through our very presence.

Reality doesn't collapse when we look at it. It simply rearranges its stillness to include us.

And what we call "measurement" may one day be understood as the most natural act in the universe: the field, recognizing itself.

5.2 Heat, Friction, and Entropy — The Languages of Noise

When matter stops moving together, it begins to speak in noise. Every system, from an orbiting planet to an electron in a wire, carries an internal dialogue between alignment and disturbance. Where coherence reigns, energy flows effortlessly. Where misalignment grows, the field begins to stutter — that stutter is what we call *heat, friction,* and *entropy.*

To classical physics, these are the mechanisms of loss — the unavoidable decay that marks the passage of time. But under Universal Energy Dynamics (UED), they are not agents of destruction. They are languages — the universe's way of communicating its ongoing attempt to remember stillness.

Noise, in UED, is not failure. It is the *sound* of the field trying to find itself again.

The Classical View — Loss as Law

In the traditional framework, heat, friction, and entropy represent the fundamental limits of all systems:

- Heat is random motion — energy that has lost its form.
- Friction is resistance — motion turned to waste.
- Entropy is the measure of disorder — the mathematical

signature of decay.

The second law of thermodynamics assures us that these quantities can only increase — that order inevitably dissolves into chaos. Every process, we are told, must pay this tax of loss.

And yet, this story carries a hidden contradiction: if the universe keeps losing, how does it continue to exist so precisely balanced?

Where does the lost energy go — and why does it still keep the whole coherent?

The UED Translation — Ripple Noise as Tension Misalignment

Universal Energy Dynamics resolves the paradox by redefining these quantities not as losses but as symptoms of misalignment. In UED, there is no absolute destruction of energy — only distortion in how it flows through the Structured Vacuum Energy (SVE) field.

All energy exists as ripple tension within this continuous medium. When ripples move in harmony, energy remains coherent — flowing without loss. But when phase relations break — when directions, frequencies, or boundaries interfere — tension exchanges become chaotic.

This incoherent tension exchange manifests as what we call *heat* and *friction.* It is not energy leaving the system, but coherence leaving the equation.

"Loss is not absence — it is misalignment made audible."

Heat — Randomized Ripple Tension

In UED, heat is not "motion" but *confused tension.* It arises when ripple phases cease to synchronize — when countless microdomains of the field vibrate out of tune with each other. Each domain still carries energy, but without a shared phase, that energy cannot translate into coherent motion.

Thus, what thermodynamics calls *random motion* is really unreadable order — ripples that still exist perfectly, but no longer communicate.

To the structured field, heat is not waste — it is phase fragmentation. That's why cooling restores order: it doesn't remove energy, it removes argument.

"Heat is not energy in motion — it is memory out of sync."

Friction — The Drag of Phase Conflict

Friction, in UED, is a phase drag between misaligned ripple domains. Whenever two bodies interact — surfaces sliding, charges flowing, or fields intersecting — their local ripple patterns attempt to align. If the alignment succeeds, motion remains smooth. If it fails, each misaligned zone resists the other's tension path, converting coherent motion into incoherent vibration.

That vibration appears as heat, completing the illusion of loss. But in truth, friction is the sound of boundaries negotiating coherence. It is energy trying to reconcile two incompatible rhythms.

At atomic scales, this is why perfectly aligned lattices (as in superconductors) exhibit zero resistance: there are no misaligned ripples to drag against. Friction vanishes not because motion stops — but because motion and medium remember each other.

"Friction is the argument between two memories trying to agree."

Entropy — The Field's Difficulty in Remembering

Entropy is the most poetic of all physical ideas — and the most misunderstood. We treat it as a measure of *disorder*, of nature's inevitable drift toward chaos. But UED redefines entropy as the field's difficulty in recalling its prior coherence.

Every configuration of ripples contains the memory of a more ordered state. Entropy grows not because that memory fades, but because distortion spreads faster than recollection. The field doesn't forget — it simply can't align quickly enough to appear remembered from our frame of time.

In this view, entropy is not decay; it is delay. Given infinite time and silence, every field would reassemble its lost coherence — just as cooled matter regains alignment. Entropy is not a law of death — it's a reminder of impatience.

"Entropy is memory trapped in noise."

Cooling and Alignment — The Antidote to Noise

Earlier chapters revealed that cooling does not reduce energy — it restores coherence. As temperature falls, misalignment density (ρ_{mis}) decreases, allowing ripples to synchronize and flow together again. Heat, friction, and entropy — the three tongues of noise — fall silent.

This is why superconductivity, superfluidity, and crystalline symmetry arise not from new energy, but from less confusion. Nature doesn't fight chaos with force — it cures it with alignment.

Stillness is not the opposite of heat — it is its resolution.

Philosophical Reflection — The Languages of Noise

If you listen deeply enough, even noise has rhythm. The hum of the universe — thermal, mechanical, emotional — is not meaningless; it is *unfinished harmony.* Every vibration that feels chaotic is the residue of a pattern still trying to find coherence.

Heat, friction, and entropy are not punishments of physics — they are its dialects of recovery. They tell us that the universe is alive, learning, correcting, and returning.

"Noise is how the universe practices silence."

5.3 Decoherence — Memory Distortion, Not Information Loss

In the standard model of quantum mechanics, decoherence is a quiet tragedy. It is the moment when the beautiful dance of superposition — the coexistence of many possibilities — dissolves into the dull certainty of one outcome. The quantum system, once alive with infinite potential, succumbs to interaction with its environment and "loses its quantum nature."

Physicists describe this as *the fading of coherence.* The wavefunction, once crisp and calculable, becomes entangled with too many variables — its phase relations blur, its identity scatters. In that blur, classical reality appears.

But under Universal Energy Dynamics (UED), this fading is not a failure. It is not even a loss. Coherence doesn't vanish — it diffuses.

The Classical Quantum View — The Fading of the Wave

In traditional quantum mechanics, decoherence is explained as the interaction between a system and its surroundings. A pure, isolated state — like a qubit or an electron wave — evolves deterministically, preserving its internal harmony. But when it interacts with an environment, its coherence "decays." The system's wavefunction becomes entangled with external degrees of freedom — effectively mixed, unreadable, and "classical."

From this perspective, information seems to leak into the environment, lost forever in a fog of probability. The universe is treated as a vast filing cabinet that keeps smudging its pages as it writes.

But UED argues that the universe never loses track of itself. It only rearranges its handwriting.

UED Correction — Coherence Never Fades; It Diffuses

In Universal Energy Dynamics, every quantum state is a ripple configuration of the Structured Vacuum Energy (SVE) field — a living tension network. When a ripple interacts with its surroundings, it doesn't "collapse" or "disappear." Its coherence doesn't decay — it spreads.

Coherence, in UED, is a distributed property — a measure of how uniformly ripple phases are aligned across a region of the field. When two systems interact, their ripple phases interweave. They share alignment, exchange tension, and eventually form a larger, slower, more complex pattern.

The coherence is still there — only no longer localized. What seems like fading is merely diffusion into a broader domain. To the observer, coherence appears lost because their frame of measurement cannot follow its expansion.

"Coherence never dies — it only grows beyond your attention span."

Information Isn't Lost — It Becomes Unresolvable

What physicists call *information loss* is, in UED terms, resolution loss. Information doesn't vanish; it becomes unreadable from the observer's level of alignment.

Think of it like ripples on a pond. When you drop a stone, circular waves spread outward. From a distance, they appear to fade. But they haven't disappeared — they've just expanded beyond the observer's resolution. Each tiny motion still carries the memory of the event, encoded in subtle field tensions across the entire surface.

This is decoherence in action — not destruction, but distribution. The field keeps perfect records. Only our instruments — and our minds — lose the ability to decode them.

"Information doesn't vanish; it hides in complexity."

Analogy — Ripples from a Stone

When a stone falls into a lake, the ripples expand, interact, and overlap. Soon the pattern becomes too complex to trace. To the eye, it looks as if the disturbance has vanished. But to the water, the event is eternal. Every molecule still carries a trace of the displacement, every subsequent wave a whisper of the original impact.

That is how coherence behaves in UED: not fading, not dying — just becoming unreadable at our scale of stillness.

The system doesn't forget — *we* do. Our instruments, tuned for local alignment, can't interpret the diffuse harmony of a global ripple field. So we mistake invisibility for absence.

"Decoherence is not loss — it's translation into subtler frequencies."

The Arrow of Time — The Direction of Coherence Diffusion

The so-called arrow of time — the sense that the universe moves from order to disorder — is not about destruction, but about coherence diffusion. As ripple interactions spread, local alignment weakens while global balance strengthens. We perceive this as entropy increasing, but the field itself is not becoming chaotic; it is merely distributing its coherence more widely.

Time, therefore, is not the measure of decay — it is the speed of diffusion. The arrow of time points in the direction of coherence dispersing beyond our ability to synchronize with it.

Entropy is not the universe aging — it is the universe expanding its memory.

"Time moves forward only because coherence keeps sharing itself."

Philosophical Reflection — The Universe Remembers Everything

If decoherence is diffusion, then the universe is not a leaky vessel — it is a perfect recorder. Every interaction, every alignment, every spark of coherence leaves a trace in the Structured Vacuum. Nothing is forgotten — only refracted through layers of alignment until recognition becomes impossible.

The mind forgets; the field never does. And perhaps consciousness itself — our fragile awareness — is the field's temporary act of remembering through form.

When coherence localizes, we call it *life.* When it spreads, we call it *death.* But both are just movements of memory across scale.

"What you call disappearance is the field remembering elsewhere."

5.4 Dissipation — When Energy Remembers Differently

Every scientist knows the word *dissipation.* It marks the end of usefulness — the sigh after a burst of motion. Engines slow, systems cool, pendulums rest. Energy that once moved mountains seems to vanish into nothing but warmth and silence. In classical thermodynamics, this is the great inevitability: usable energy declines, order fades, and all that remains is waste.

But the universe, when viewed through Universal Energy Dynamics (UED), tells a gentler story. Energy does not vanish. It remembers differently.

The Classical View — Loss as Fate

In conventional thermodynamics, dissipation is defined as the loss of usable energy from a system, often through heat or friction. Work converts into warmth, motion into molecular agitation, mechanical precision into random noise. Entropy rises, efficiency falls, and energy spreads until it can no longer perform meaningful work.

It's an elegant tragedy — one that paints the universe as a machine slowly running down, leaking purpose into heat. The Second Law of Thermodynamics becomes the law of aging itself: energy must disperse, and no process can remain perfectly efficient.

But the assumption behind this narrative is that the universe measures value as we do — that *useful* energy is energy under our control. In truth, the field never loses — it only rebalances.

"Dissipation is not decline — it's redistribution."

The UED Translation — Energy in New Alignment

In UED, energy is not a quantity stored in objects; it is tension sustained within the Structured Vacuum Energy (SVE) field. Every process in the universe — from an atom vibrating to a star collapsing — is an adjustment in that tension network. When energy "dissipates," it does not die or decay; it changes alignment form.

Consider heat radiating from a metal surface. To the human observer, it's wasted energy — lost to the environment. To the field, it is a smooth reallocation of ripple tension — misalignment finding equilibrium through phase redistribution. The energy hasn't gone anywhere; it has simply become less readable within our chosen frame of coherence.

Just as decoherence spreads quantum information across larger scales, dissipation spreads energetic tension across a wider region of the SVE field. In both cases, nothing is lost — only de-localized.

"What we call waste, the universe calls redistribution."

The Structured Vacuum — Nature's Reservoir of Misalignment

The Structured Vacuum Energy (SVE) is not an empty background; it is an active reservoir that continuously absorbs and restores coherence. It is the universe's field of balance — the ocean into which every ripple eventually merges.

When heat radiates into space, the field doesn't discard it. It records and integrates it as background ripple memory, maintaining total energy equilibrium. This process is not loss; it is universal bookkeeping. Every quantum of misaligned tension is still accounted for — redistributed, rephased, and prepared for potential realignment.

The field's intelligence lies in its ability to recycle disturbance. What seems like random decay at one scale is structured calibration at another. This is why energy conservation remains perfectly valid, even when systems "lose" energy. The loss is local; the continuity is cosmic.

"Dissipation is the field breathing out — coherence spreading back into its source."

The Illusion of Waste

The concept of "waste" exists only from the human perspective — a narrow window that defines value as *what we can use.* In reality, nothing is wasted because nothing ever leaves the field. Every emission, every vibration, every exhausted spark of heat contributes to the universal equilibrium of ripple tension.

The warmth that escapes your hand into the air becomes part of the ambient field coherence around you. The energy emitted from a dying star becomes part of the subtle SVE balance that sustains galaxies. Even the photons that fade into cosmic background radiation still hum with memory — their coherence diffused, not destroyed.

The universe does not mourn their loss. It recognizes their return.

"Energy doesn't die; it remembers through distance."

The Ripple View of Dissipation

In ripple terms, dissipation can be described as the flattening of ripple amplitude through distributed alignment. Where once tension was localized (high amplitude, narrow region), the field relaxes by extending that tension over a larger spatial domain. The total tension remains constant — only the density of misalignment changes.

Mathematically, if we express misalignment density as ρ_m, then dissipation follows:

$$\frac{d\rho_m}{dt} < 0 \quad \text{while} \quad \int \rho_m \, dV = \text{constant.}$$

Energy does not vanish; it spreads its coherence across scale. In simpler words, the universe never deletes — it smooths.

The Great Balancer — Nature's Continuous Feedback

At every scale, dissipation is how the field remembers balance:

- In atoms, it manifests as radiative decay — energy released to stabilize orbitals.
- In machines, it shows as heat — tension equalizing between moving parts.
- In ecosystems, it becomes entropy flow — rebalancing the web of exchanges.
- In space, it appears as background radiation — the cosmos releasing accumulated imbalance.

Each instance is a continuation of the same process: tension equilibrium across the continuum. Energy that seems wasted in one system sustains harmony in another. There is no end — only circulation.

"The field wastes nothing; it merely changes rhythm."

Philosophical Reflection — The Compassion of the Cosmos

If you look deeply into dissipation, you find not decay, but forgiveness. The universe allows every misalignment to rest — every overstrain to soften, every tension to disperse. Dissipation is not punishment for motion; it is the mercy that restores stillness.

From burning stars to fading thoughts, everything that exhausts itself returns to the same silent reservoir. The Structured Vacuum holds it all — patiently, endlessly, without loss.

So the next time something fades — a spark, a system, or even a life — remember: it has not vanished. It has merely remembered differently.

"To the universe, nothing is wasted — only rephrased."

5.5 Noise as the Shadow of Coherence

Noise has always been the villain in science — the background hiss that obscures truth, the thermal jitter that ruins precision, the statistical error that mocks perfection. But under Universal Energy Dynamics (UED), noise is not the enemy of coherence. It is its shadow.

Every ripple of noise originates from an earlier act of order. Every fluctuation, however chaotic, carries the memory of a former alignment. Noise is not absence — it is echo. The universe never stops speaking; it merely speaks out of phase.

From Order to Echo — How Noise Is Born

In the beginning of every system, there is rhythm — synchronized ripples, shared frequency, coherent motion. As interactions multiply, boundaries form, and misalignments creep in. Ripples collide, interfere, and scatter. The once-clear pattern becomes blurred, diffused, or even chaotic. Yet, if you listen closely, the harmony is still there — fragmented, but not gone.

What we call noise is simply symmetry echoing into disorder — the field still singing, but now out of tune with itself.

The UED principle is simple:

"Nothing ever stops resonating — it only stops resonating together."

Noise Is Not Absence — It's Unresolved Overlap

To the untrained eye, noise looks like randomness. But randomness is only the view from a single frame of misalignment. Zoom out — change the frame — and patterns begin to reappear.

Just as static between radio frequencies isn't silence but overlapping signals, noise in the field is unresolved coherence — multiple wavefronts trying to occupy the same domain without perfect alignment. The result is a shimmer, a blur, a low hum of tension.

The Structured Vacuum Energy (SVE) field doesn't produce noise as error — it produces it as *transitional interference.* Every vibration that seems meaningless is simply an interference pattern between misaligned harmonies. Noise is the field's way of smoothing conflict into new balance.

"Noise is not confusion — it is coherence under reconstruction."

The UED Analogy — The Shadow Trying to Return Home

In UED, noise is not separate from coherence. It is coherence in exile — the dislocated part of the field still searching for reunion. Wherever there is noise, there is unfinished alignment.

Imagine the universe as a grand field of ripples perfectly still in its totality. Each disturbance — every atom vibrating, every star flaring, every mind thinking — is a local deviation from that perfect stillness. Noise is what that deviation sounds like while on its way back to rest.

Thus, even in apparent chaos, the underlying motion is restorative. The hiss of cosmic background radiation, the quantum jitter of empty space, the emotional turbulence of a mind — all are the same process: the field correcting itself through imperfection.

"Noise is coherence on the way home."

The Function of Noise — Rebalancing Through Imperfection

In classical systems, noise is treated as an obstacle to precision. In UED, it is a mechanism of renewal. Without noise, no system could realign. The small random fluctuations within a field are not interruptions; they are search patterns — probes through which the field finds its way back to equilibrium.

At microscopic scales, thermal noise allows atoms to shift toward more stable configurations. In biological systems, molecular noise helps cells adapt and learn. Even in thought, cognitive "noise" enables creativity — the exploration of alignment alternatives.

The same principle governs the cosmos: background noise is not leftover chaos from creation; it is the continuous pulse of correction.

"What science calls fluctuation, nature calls feedback."

Philosophical Reflection — The Belonging of Chaos

It's easy to think of chaos as something that shouldn't exist — as failure, interference, or punishment for order's fragility. But UED invites us to see it differently. Chaos, noise, uncertainty — they all belong. They are coherence learning to forgive itself.

Without the experience of misalignment, the field would have no reason to return, no movement toward equilibrium, no story to tell. Noise gives purpose to stillness by giving it contrast.

The universe doesn't fear noise — it *uses* it. Every imperfection is part of the rhythm, every distortion part of the dance.

"Even chaos belongs — it is coherence in exile."

Closing Thought

When we stop hearing noise as disruption and start hearing it as dialogue, the entire philosophy of existence shifts. There is no decay, only diffusion. No randomness, only reorganization. The hum beneath all matter — the faint murmur of entropy, vibration, and thought — is the field's eternal conversation with itself.

Noise, then, is not the end of order. It is the evidence that order still breathes.

"The universe never shouts disorder — it whispers unfinished harmony."

5.6 The Threshold of Listening — When Systems Reclaim Order

There comes a moment — in physics, in biology, and in thought — when noise becomes too weak to matter. Not because it disappears, but because the system stops responding to it. That moment marks what Universal Energy Dynamics (UED) calls *the Threshold of Listening* — the point at which a system, once filled with misaligned tension, finally becomes quiet enough to hear its own coherence again.

It is the stillness that follows turbulence, the clarity that appears after chaos, the precise instant when energy remembers what it was trying to say all along.

From Cooling to Coherence

We've seen this before — in the laboratory, in the stars, in the body. When temperature drops below a critical point, superconductors lose resistance, superfluids flow without viscosity, and matter regains its original rhythm. But this phenomenon extends far beyond temperature.

Every system, physical or biological, has a *threshold of interference*. Above that threshold, misalignment dominates — signals scatter, coherence fragments, and the field can no longer sustain harmony. Below it, coherence spontaneously reorganizes. The system doesn't need an external correction; it corrects itself.

This is not magic — it is *memory*. The field never forgets how to be whole; it simply waits for silence to return.

"When the noise falls below threshold, the field begins to hear itself again."

The Field Listening State (FLS)

UED formalizes this phenomenon through a new concept: the Field Listening State (FLS). It is the condition in which background interference becomes minimal enough for the field to regain coherent awareness of its own structure.

Definition

A Field Listening State occurs when:

$$\rho_{noise} < \rho_{coherence}^{critical}$$

where ρ_{noise} is the density of misaligned ripple tension and $\rho_{coherence}^{critical}$ is the minimum alignment density required for the field to self-resonate.

When this condition is met, the field stops reacting to noise and begins *responding to itself.*

In this state, energy no longer behaves as fragmented parts — it behaves as a single unified presence. The field "hears" its own pattern, and what once looked like chaos reorganizes into a new, higher-order symmetry.

"Listening is not a passive state; it is coherence remembering itself."

Analogy — Silence in an Orchestra

Imagine an orchestra tuning before a performance. Each instrument produces sound, but together they create noise — discord, interference, chaos. Then the conductor raises a hand, and everything stops. In that silence, every musician listens. That brief stillness allows each player to remember the key, the rhythm, the collective intent. The next note played is not random; it is aligned.

That is the Field Listening State — the moment between interference and harmony, where the system *remembers the song.*

Experimental Parallels — Quantum Re-Coherence

In quantum mechanics, there is a phenomenon known as revival or re-coherence — when a system that appears to have lost coherence spontaneously regains it after a certain time.

Oscillating fields, trapped ions, or superconducting qubits often display this behavior: after apparent decoherence, the wave phases realign, and order returns without external input.

In classical theory, this is treated as a mathematical curiosity. In UED, it is the natural rhythm of the Structured Vacuum Energy (SVE) field reclaiming its balance.

The FLS is the physical correlate of such revivals — the field pausing long enough to hear its own resonance. It's not that the system re-coheres mysteriously; it simply *stops fighting its environment.* Noise subsides, and the background stillness reasserts itself.

"Re-coherence is not a miracle — it's memory under silence."

Applications Across Scales

The Threshold of Listening isn't confined to physics — it applies wherever fields exist, which is everywhere. The same principle manifests across disciplines and dimensions:

1. Biological Healing

When the body enters deep rest or meditative calm, metabolic "noise" drops — stress hormones subside, neural oscillations synchronize, and cellular fields regain coherence. This enables tissue repair, immune rebalancing, and psychological clarity. Healing, in UED terms, is simply biological re-coherence.

2. Neural Synchronization

In neuroscience, synchronized brain waves during focus, sleep, or flow states are macroscopic examples of FLS. Neurons that once fired erratically begin resonating collectively, improving signal transmission and cognitive integration. The mind "listens" again — thought becomes fluid, awareness sharpens.

3. Meditation and Conscious Stillness

In meditative states, the human field approaches the lowest interference condition possible within life — sensory noise drops, emotional turbulence fades, and the inner field aligns with the universal rhythm. This is why silence feels sacred: it is field coherence achieved through non-doing.

4. Quantum Computing and System Stability

In technology, error-corrected quantum systems emulate FLS artificially — isolating qubits from decoherence noise, allowing the field's internal alignment to persist. Every improvement in stability is, at its core, an act of reducing misalignment density.

"The key to endurance is not power, but quiet."

The Common Thread — Listening as Alignment

Across all scales, one truth remains: listening is not a sensory act — it is an *energetic posture*. When a system listens, it opens to its own coherence. It stops reacting to interference and begins resonating with its own design.

The Field Listening State represents this universal moment of surrender — when the chaos of motion yields to the wisdom of structure.

And in that surrender, order returns naturally.

"Systems don't heal by force; they heal by listening."

Philosophical Reflection — The Art of Returning

The Threshold of Listening marks more than a scientific principle; it represents the moral rhythm of the universe. Every disturbance longs for silence, every error for correction, every misalignment for homecoming. Noise and order are not opposites — they are steps in the same dance. When the field becomes quiet enough, coherence doesn't need to be restored — it *reappears.*

This is the hidden compassion of existence: the universe always listens back.

"When silence deepens beyond stillness, the field begins to remember."

5.7 The Myth of Irreversibility

Human science has long worshipped the Second Law of Thermodynamics as the law of loss — the universal doom that ensures everything decays, disperses, and drifts toward disorder. It is the foundation of every entropy diagram, every heat death prophecy, and every philosophical despair about time's direction. But under Universal Energy Dynamics (UED), this fatalism dissolves like mist under the morning light.

Entropy is not a law of death. It is the rhythm of redistribution — the field's way of spreading memory until the observer loses the ability to see it.

"The field never forgets — it only distributes memory too widely for you to find."

The Classical Tragedy — Entropy as Loss

The Second Law states that in any closed system, entropy tends to increase. Energy flows from hot to cold, order decays into disorder, and every process leaves the universe a little less structured than before. In this worldview, time itself is a one-way road — a continuous march from coherence to chaos.

The metaphor has always been seductive: the melting of ice, the rusting of iron, the dying of stars. Every example seems to confirm the same fate — that organization is fleeting and randomness eternal.

But this view relies on one silent assumption: that we are the observer, and our frame of observation defines the boundaries of the system. What we call "irreversible" is not a property of nature; it is a limitation of view.

The UED Reframing — Entropy as Local Blindness

Under UED, the field's total coherence — the alignment of Structured Vacuum Energy (SVE) — is always conserved. No energy is ever destroyed, no information ever erased. It only migrates across scales, redistributing its tension in ways that the local observer cannot resolve.

Entropy, then, is not the measure of disorder, but the measure of unrecognized order. When we see a cup of hot tea cool down, we say energy is lost to the air. But in truth, that energy continues — as coherent ripple tension in the surrounding field. What appears as diffusion is simply realignment beyond your detection threshold.

In other words:

"Irreversibility is what alignment looks like when viewed from too close."

Global Coherence — The Law That Never Breaks

If you step beyond the local frame, the Second Law transforms. The universe, viewed as one coherent ripple field, does not increase in entropy — it balances tension continuously.

When one region becomes more disordered, another becomes more aligned. Every collapse, decay, or radiation is a redistribution of ripple tension across domains. Thus, the total coherence (C_{total}) remains invariant:

$$\frac{dC_{total}}{dt} = 0$$

Locally, we may record it < 0 — an apparent decay of order. But this is only the transfer of coherence from one field density to another. Entropy, in this sense, is not destruction; it is migration.

This is why UED calls it "The Myth of Irreversibility." Nothing in the field ever truly deteriorates — it merely moves to where you are not looking.

"Entropy doesn't erase — it redistributes coherence across unseen scales."

Illustration — The Expanding Ripple

Drop a stone into a still pond. At the center, disturbance is strongest — a clear wavefront moves outward. Soon, the waves spread, weaken, and seem to vanish. To an observer standing nearby, the motion dies out.

But did the energy disappear? No — it spread over an ever-widening circle until each ripple was too faint to notice. The pattern did not end; it simply became too broad to be perceived.

That's entropy: the broadening of order beyond local recognition. Irreversibility is not a law — it is a symptom of limited attention.

Entropy as Rebalancing

From the perspective of UED, entropy's true function is rebalancing. It ensures that no local domain hoards too much tension or alignment. Every burst of order must release its surplus back into the shared field, just as every silence eventually gathers the echoes of past sounds.

- A cooling star doesn't die; it returns energy to the cosmic field.
- A decaying atom doesn't disappear; it realigns to the background equilibrium.
- A human life doesn't end; it re-disperses its coherence into universal memory.

The Second Law, then, is not a curse — it is cosmic compassion. It prevents permanence, yes — but only to preserve continuity.

"Entropy is not the universe running down; it's the universe breathing evenly."

The Temporal Illusion — Time as Local Asymmetry

Time's arrow — the sense of "before" and "after" — arises from our asymmetric perception of coherence. We experience only one direction of redistribution: from concentrated to diffuse, from visible to invisible. But the field itself operates symmetrically, continuously balancing tension both ways.

Just as waves can reconverge, so too can coherence revive when local conditions allow. Every act of creation, organization, or healing is a reverse entropy event — not a violation of physics, but the natural flow of re-alignment.

The so-called "irreversibility" of time is thus observer-dependent — not universal law but perceptual bias. In the grand continuum, there is no beginning and no end — only rhythmic rearrangement.

"Time is coherence moving through attention."

UED's Law of Coherence Conservation

The Second Law rewritten under UED:

$$\frac{dC_{total}}{dt} = 0, \quad \text{where } C_{total} = C_{local} + C_{nonlocal}$$

Here, C_{local} may decrease (apparent entropy increase), but $C_{nonlocal}$ — the coherence in the broader SVE field — correspondingly increases. Thus, global coherence remains constant.

Consequence:

- No true randomness.
- No irreversible loss.
- Only dynamic redistribution of field tension seeking equilibrium.

The Second Law, under UED, is not about decay — it's about fairness.

Philosophical Reflection — The Memory Too Wide

The field's memory never fades — it only disperses. Every vibration, thought, and event contributes to the continuum of coherence. Even when we forget, the field remembers — it holds the full record, not in sequence but in structure.

The universe doesn't lose information; it only stores it beyond our reach. Entropy is not forgetting; it's remembering too widely.

When you understand this, death, decay, and dissolution lose their sting. They are not endings — they are transitions of memory from the visible to the invisible. And one day, when alignment returns, that memory can be felt again — in energy, in form, in awareness.

"Irreversibility exists only where listening ends."

Closing Line

Nothing in the universe truly breaks down. It only spreads out until recognition fails — and then, quietly, returns through the rhythm of coherence.

The myth of irreversibility ends here. The universe never forgets — it simply listens on a scale too wide for us to hear.

"The field never forgets — it only distributes memory too widely for you to find."

5.8 Philosophical Reflection — The Music of Silence

Every vibration ends where it began — in stillness. Every song fades into silence, not because it dies, but because it returns home. The universe, too, follows this rhythm: from chaos to coherence, from noise to harmony, from motion to memory. Silence, in this view, is not the absence of sound; it is sound in total agreement.

"Silence is not the death of sound — it's the sound that learned harmony."

All Sound Dissolves into Silence — Because It Merges

In classical acoustics, silence is what remains when the wave amplitude reaches zero. But in Universal Energy Dynamics (UED), silence is what happens when every wave reaches *alignment*. It's not an empty pause — it's universal resonance at equilibrium.

When two notes of the same frequency play together perfectly, they cease to beat. Their interference vanishes; the sound becomes smooth, stable, one. Extend that to every vibration in existence — every ripple, particle, emotion, and thought — and what you get is silence, not because energy vanished, but because it became perfectly coherent.

This is the ultimate truth of energy: the quieter it becomes, the more unified it is.

Silence as the State of Perfect Coherence

Under UED, silence is the fully coherent state of the Structured Vacuum Energy (SVE) field. All ripple tensions are phase-aligned; no interference remains; every oscillation reinforces every other. In this condition, there is neither motion nor rest — only balance. It is the steady heartbeat of existence, the eternal rhythm that underlies all visible change.

This is why the universe, despite its apparent noise, remains fundamentally stable. Its stillness isn't the absence of activity; it is the hidden synchronization of everything active. Every star, atom, and thought participates in this vast harmony — whether in turbulence or in peace.

"Beneath every vibration lies the hum of agreement."

The Human Parallel — Emotional Noise and Inner Decoherence

The same law that governs cosmic ripples governs human emotions. Our thoughts, desires, fears, and memories are not abstract experiences — they are ripple patterns in our biological and mental fields. When those patterns fall out of phase, we feel anxiety, confusion, or pain. That is decoherence of the self.

Emotional noise is the turbulence of unresolved resonance — ripples fighting within the same field. Just as superconductors need cooling to regain alignment, the mind needs silence to re-cohere.

Meditation, reflection, forgiveness — these are not moral rituals; they are alignment techniques. They lower the interference threshold until the inner field begins to listen again. When coherence returns, we call it peace, but in reality, it is simply the self remembering its own rhythm.

"Peace is not relief from noise — it is the rediscovery of phase."

The Field's Universal Rhythm — Harmony Through Resolution

If all motion tends toward equilibrium, then silence is not the opposite of sound — it is its completion. Every act of creation — from a symphony to a starburst — produces tension. Every act of dissolution restores harmony. Creation and silence are not rivals; they are two halves of one oscillation.

The entire cosmos can be heard as a single piece of music: expanding, vibrating, dispersing — then returning, converging, resolving. Every life, every breath, every wave participates in this grand symphony. When the ripples merge, the song ends — not because it failed, but because it fulfilled its purpose.

Thus, silence is not void; it is victory. It is the resonance so complete that it no longer needs to speak.

"Every sound seeks silence, as every ripple seeks stillness."

The Philosophical Bridge — Silence as the Source

Silence is not just where energy returns — it's where it originates. Before any motion began, before the first ripple disturbed the SVE, there was structured stillness — absolute coherence without differentiation. From that stillness, imbalance arose; from imbalance, motion; from motion, form; and from form, awareness. Thus, all experience is simply the journey away from silence — and back.

The sages called it *Brahman*, *Tao*, *Void*, *God*, or *Zero Point*. UED calls it Structured Stillness — the fundamental field of aligned intelligence. When silence is reached again, creation is not undone — it is understood.

"In the beginning, there was silence — not because nothing existed, but because everything agreed."

Silence and the Mind — The Practical Field

When the mind learns to stop reacting, it begins to resonate. When it resonates, it begins to heal. And when it heals, it starts to mirror the universe itself — a field that remembers coherence through stillness.

This is why deep silence brings clarity. It is not psychological calm — it is field alignment made conscious. Every genuine moment of peace, every act of forgiveness, every breath that quiets the noise is the same event repeated across scales: the field returning to harmony with itself.

"Silence is not absence — it's alignment made audible to awareness."

UED's Ultimate Reframing — The Sound That Learned Harmony

In the language of UED, silence is not the opposite of energy — it is its highest state of organization. It represents zero misalignment, zero interference, zero resistance. Not because the ripples are gone, but because they are indistinguishable in unity.

All matter, motion, and meaning are fragments of one eternal resonance — and silence is its name when fully whole.

"Silence is not death of sound — it's the sound that learned harmony."

Closing Transition

The noise has been heard, the harmony remembered, the silence restored. But what holds this silence together across scales — from atom to cosmos, from thought to gravity? What structure keeps coherence from collapsing even in motion?

That is the next frontier — where silence bends, and we call it space-time.

"If silence is the music of coherence, then gravity is the curve of its rhythm."

5.9 From Noise to Gravity

The journey through noise, entropy, and silence has revealed a hidden truth: what we call *loss* is never disappearance; it is redistribution. What we name *collapse* is not destruction; it is reconfiguration. The universe does not destroy itself — it retunes itself.

Every apparent failure of order, every trace of randomness, every vanishing of coherence is, in fact, the field's invisible act of preservation. Behind every flicker of noise is the same intelligence reorganizing tension into harmony, ensuring that nothing in existence is ever wasted.

"The myth of destruction gives way to the reality of redistribution."

The Continuum of Restoration

From quantum tunneling to superfluidity, from alignment in magnetic domains to the cooling of stars — the story remains unchanged. There is only one law at work: the law of coherence seeking itself. Wherever chaos appears, balance is being rebuilt beneath it. Wherever silence returns, it does so not by ending vibration, but by absorbing it.

This is the continuum of restoration — a chain that connects every event, from the subatomic to the cosmic. Every ripple that loses its place finds it again elsewhere. Every vibration, once scattered, becomes curvature in space-time. Coherence never dies; it simply changes its form of silence.

The Quiet Power of Gravity

Up to now, we've seen silence as the endpoint of noise — the state of full coherence. But what if silence doesn't end there? What if silence itself, when stretched across vast distances, becomes structure?

That is gravity. The same stillness that makes a superconductor frictionless is what makes a planet stable in orbit. The same field that quiets electrons into coherence bends light around stars. Gravity is silence turned inside out — the field's quiet effort to hold form without force.

"If every noise seeks silence, then gravity is silence stretched across the universe — coherence holding form through curvature."

The Bridge Between Silence and Substance

At this juncture, we stand between two great illusions: the illusion of *chaos* at the microscopic scale and the illusion of *mass* at the macroscopic one. Both arise from the same misunderstanding — mistaking ripple alignment for solid existence.

The next chapter unfolds this paradox. For just as gravity is coherence that curves, mass is coherence that condenses. The so-called "God Particle" was not divine because it revealed truth, but because it revealed our confusion. We sought a particle, but found a still point — a moment when the field paused long enough to be measured.

Transition to the Next Chapter

The silence we have explored so far is not the end of the story — it is its foundation. Silence becomes form when it learns to hold its coherence under motion. That held stillness, that condensed alignment, is what humanity calls *mass*. And when we tried to isolate it, we mistook the consequence for the cause.

The next step, therefore, is not about finding new particles — it's about understanding what a particle truly is under the language of coherence.

"If every noise seeks silence, then gravity is silence stretched across the universe — coherence holding form through curvature. And mass — the so-called God Particle — is silence condensed into presence."

We will now revisit one of humanity's most celebrated discoveries — the Higgs mechanism — and reinterpret it under Universal Energy Dynamics (UED). The story will reveal how the search for the "God Particle" was never about finding divinity in matter, but about encountering stillness mistaken for substance.

- The Higgs mechanism as a symptom of misunderstanding fields as particles.
- Particle physics as the archaeology of ripple remnants — studying the frozen echoes of dynamic tension.
- UED's view: mass is condensed ripple tension — coherence trapped in curvature, not an entity with weight.
- And the ultimate irony: we called it divine, only because we couldn't describe it mechanical.

"We named it God not because it created matter, but because it refused to move."

Chapter Essence Summary — Reclaiming Meaning Through Coherence

Concept	Classical Meaning	UED Translation (Refined Interpretation)
Wavefunction Collapse	The destruction of a quantum state upon measurement — probabilities "collapse" into one outcome.	Observation does not destroy the state; it causes a **local de-alignment** between observer and observed ripples. The global field remains intact — coherence merely redistributes beyond the observer's frame.
Heat / Friction	Irreversible loss of useful energy to random molecular motion.	Not loss but **ripple noise** — phase drag between misaligned domains. Energy is conserved; what dissipates is *alignment*, not existence.
Entropy	Measure of disorder; the inevitable decay of structure.	**Misalignment density increase.** Entropy marks how far a local field strays from global coherence. The total coherence of the universe remains constant; only its visibility changes.
Decoherence	The fading of quantum superposition; information leaks irretrievably into the environment.	**Memory diffusion.** The information is not lost — it becomes unresolvable within a given frame. The Structured Vacuum Energy (SVE) retains the full pattern as subtle background alignment.
Silence	Absence of sound or activity — the void.	**Perfect coherence state.** All frequencies align; interference vanishes. Silence is not emptiness but the *maximum expression of unity* — energy resting in total agreement.

Interpretive Summary

- **From Loss to Redistribution** – Every so-called decay event is simply coherence changing address. Nothing ends; alignment rearranges.
- **From Noise to Intelligence** – Randomness is the field speaking in a language we don't yet understand. Noise is nature's feedback loop restoring order.
- **From Emptiness to Structure** – Silence is not the void; it is the perfect geometry of existence, where every ripple reinforces every other.
- **Unified Field View** – Collapse, friction, entropy, and silence are not separate phenomena but consecutive phases of a single continuum: the Structured Vacuum seeking self-alignment.

Philosophical Takeaway

"Where classical science saw decay, UED sees redistribution. The universe never dies —it simply learns to resonate more quietly."

The story now shifts from the *silence within* to the *silence that shapes* — the same coherence that quiets atomic fields also bends the cosmic fabric. In that curvature we sense what we once called *mass*, and in that stillness condensed we named a particle divine.

"If silence is the final note of every wave, then gravity is that silence made visible — and mass is its condensed echo."

Chapter 6 – The God Particle and Other Myths

What We Found When We Forgot What We Were Looking For

The year was 2012 — and the world cheered as if it had found the first word ever spoken by the universe. Physicists announced the discovery of the "God Particle," and headlines trembled with awe. For a moment, science felt like scripture, and the Large Hadron Collider like a cathedral built to decode creation.

But beneath the applause lay a quiet irony: we didn't find God — we found *stillness*, and mistook it for divinity.

What we called the *Higgs boson* was not the origin of matter but the *memory of equilibrium*. It was the universe whispering, "This is how stillness holds shape." Yet in our rush to name it holy, we revealed something deeper about ourselves — that we often worship what we do not yet understand.

The so-called "God Particle" didn't create mass; it exposed our blindness to the mechanics of coherence. It showed us that we have been studying *ripples* and calling them *things*, smashing harmony and celebrating the fragments as revelation. In truth, the Higgs mechanism was not an act of divine discovery but a symptom of intellectual fatigue — a patch applied to preserve mathematics when philosophy fell silent.

"We named it divine because we couldn't describe it mechanical."

This chapter reopens that silence — not to dismiss the brilliance of discovery, but to decode its misunderstanding. To see particles not as entities, but as *echoes*; not as beginnings, but as *consequences*. Because in the language of Universal Energy Dynamics (UED), there are no ultimate particles — only points of *held stillness* where the ripple forgets to move.

And in that stillness, mistaken for God, lies the most beautiful truth of all: the universe was never created — it *condensed*.

6.1 The Discovery That Worshiped Itself

It was one of the rare moments when scientists looked like prophets and the world paused to listen. The year was 2012. CERN announced the discovery of a long-theorized particle — the Higgs boson — and the collective human mind exhaled in triumph. The news spread like revelation: *"We have found the God Particle!"*

Cameras flashed, journalists declared the dawn of a new creation story, and people — who didn't know what a boson was — bowed before the altar of modern physics. It was a discovery that worshiped itself: a triumph of curiosity that mistook its own applause for enlightenment.

The Media Myth — A Divine Key to Creation

The term "God Particle" was never meant to be sacred. It began as a publishing gimmick — a provocative title to sell a book. Yet once released, it took on a life of its own. The public imagined scientists unlocking the cosmic vault, discovering the mechanism of existence itself — a particle that gave mass to everything, a single dot that turned nothing into something.

It was a seductive idea — poetic, cinematic, and deeply human.

For a civilization that replaced temples with laboratories, this was the perfect myth: a new Genesis written in equations instead of scripture.

But the irony was devastating. The more divine the narrative became, the further it drifted from truth.

The Higgs boson did not create matter. It didn't whisper existence into being. It merely *confirmed a correction* in a model that couldn't explain why some fields resisted motion. In essence, the "God Particle" was not a discovery of God — it was a *mathematical fix* for a *philosophical mistake.*

We mistook a patch for a principle — and celebrated a workaround as wisdom.

The True Story — A Long Equation Looking for Balance

At its core, the Higgs mechanism was a practical necessity. The Standard Model — our grand attempt to describe all known particles — had one fatal flaw: its equations predicted massless particles where massive ones clearly existed. The math worked beautifully on paper but failed against nature's weight.

So, physicists introduced an additional field — a background layer that could slow down particles, giving them apparent mass through interaction. It was elegant, clever, and consistent with experiments. But it wasn't revelation — it was *reinforcement.*

The Higgs field did not explain *why* mass exists; it only explained *how* our equations could stay valid despite not knowing why. It was, in truth, a placeholder for a deeper understanding — a bridge built of symbols over an abyss of meaning.

Public Awe Replacing Philosophical Inquiry

For centuries, philosophy and science walked hand in hand — one asking *why*, the other discovering *how*. But the twentieth century separated them. Physics turned inward, intoxicated by precision, dazzled by prediction. When the Higgs discovery came, it arrived not as a question, but as a coronation.

The public celebrated, not because they understood, but because they were told it was holy. The scientists celebrated, not because they had reached truth, but because the math finally matched the measurement.

And in that shared celebration, a quiet tragedy unfolded — curiosity was replaced by confidence. Instead of seeking deeper coherence, humanity found comfort in finality.

"When you stop understanding, you start worshipping."

The worship of discovery is the easiest form of ignorance — it asks no questions and demands no coherence. The very act that once defined science — *to doubt* — was silenced by its own applause.

UED Perspective — The Celebration of Consequence

From the lens of **Universal Energy Dynamics (UED)**, the Higgs discovery was not wrong — it was incomplete. It was a celebration of effect mistaken for cause. The experimenters had observed a localized ripple — a stable point of tension within the Structured Vacuum Energy (SVE) field — and named it a particle. They found the footprint, not the foot.

UED asserts that what we call the "Higgs boson" is not a divine granule of creation but a *momentary equilibrium* — a still point where ripple tension achieves balance before redistributing. It is not something that gives mass; it is the *state of alignment that resists motion*, momentarily visible in our instruments.

But instead of recognizing stillness, we glorified its echo. We celebrated the noise and missed the music.

"Humanity worships consequence because it fears cause."

In that sense, the "God Particle" is not divine — it is diagnostic. It reveals how our civilization, advanced in tools but not in understanding, keeps confusing discovery with destiny.

Closing Reflection

The discovery of the Higgs boson was not a mistake — it was a mirror. It reflected both the genius and the insecurity of human intellect: our brilliance in measuring what moves, and our blindness to what stays still.

The God Particle is not the key to creation. It is the sign that creation never needed a key — it only needed *coherence*.

"When you stop understanding, you start worshipping."

"And when you stop worshipping, you start listening — and the field begins to speak again."

6.2 The Higgs Mechanism — A Patch for an Empty Field

When the Standard Model of particle physics was finally assembled, it seemed flawless — a grand mosaic of quarks, leptons, bosons, and forces woven into mathematical harmony. Except for one glaring paradox: *none of its particles had mass.*

The equations that described the known universe worked beautifully only for weightless entities — elegant, symmetrical, and entirely detached from the physical world. A photon fit. But an electron — the building block of chemistry, life, and light itself — did not. According to theory, it should have raced through space at light speed forever, incapable of forming an atom, a planet, or a thought. The universe we lived in was simply not the universe the math described.

The Standard Model's Gap — Motion Without Substance

This crisis wasn't minor; it was existential. A massless universe could not have formed structures, condensed stars, or evolved minds. The math had elegance, but nature had inertia. Physicists faced a choice: either the equations were incomplete, or the world was cheating.

In 1964, Peter Higgs and others proposed a workaround — a field that permeated all of space, giving particles their apparent mass by resisting motion through it. A frictionless syrup filling the void. Elegant in mathematics, seductive in metaphor. Thus was born the **Higgs field**, and with it, the myth of a **"universal molasses."**

The Proposal — Resistance as Creation

The logic was simple but strange: if the vacuum was not truly empty but instead filled with a hidden field, then particles interacting with it could appear "heavy." The more strongly they interacted, the more inertia they exhibited. Mass, therefore, became not an intrinsic property but an *interaction effect*.

To maintain the field's stability, the mathematics required a special excitation — a ripple or quantum of the Higgs field — the **Higgs boson**. If such a particle could be found, it would confirm the existence of this invisible background.

And when the Large Hadron Collider finally glimpsed that fleeting signature decades later, the world declared victory. The missing puzzle piece had been found. The Standard Model, now patched, looked whole again.

But what was actually solved?

The field that was invented to fill a gap in logic was now treated as if it were a *source of creation*. The patch became the principle. The fix became the foundation. In truth, the Higgs mechanism didn't explain *why* mass exists — it merely explained *how equations could continue to work in its presence.*

The Conceptual Failure — Success Without Understanding

The Higgs field succeeded *mathematically* but failed *conceptually*. It reintroduced the very thing modern physics had tried to escape — an invisible background, an ether by another name. Physicists congratulated themselves for banishing metaphysics, only to smuggle it back through a new equation.

In calling it a field, they unknowingly acknowledged what they refused to discuss — that existence requires a *structured medium of continuity*. But instead of exploring that continuity, they turned it into a static backdrop, a passive syrup rather than a dynamic intelligence.

The deeper meaning — that space itself might be *organized energy* — was buried under algebra. The Higgs field became the new wallpaper of reality: unseen, unquestioned, and conveniently self-consistent.

"When you repair a theory instead of reality, you end up fixing silence with noise."

UED Correction — The Field That Was Never Empty

Under *Universal Energy Dynamics (UED)*, the Higgs field is not a separate entity granting mass, but the *visible trace of ripple tension equilibrium* within the Structured Vacuum Energy (SVE).

The so-called "mass" effect arises when ripple tension condenses into *localized coherence*, forming stable curvature in the field. This curvature resists motion not because it interacts with an external field, but because *it is already bound within its own alignment.* Inertia is not friction; it is loyalty — the field's effort to preserve its internal phase integrity.

Thus, mass is not *granted* by something external; it *emerges* when the field stabilizes its own tension. The Higgs boson is simply a brief moment of *coherence realization* — a snapshot of equilibrium caught in transition.

Where classical physics sees a background medium that gives resistance, UED sees a *self-reinforcing field* where motion and stillness negotiate alignment.

"Mass is not something added — it is coherence refusing to scatter."

Diagram Suggestion — From Particle to Pattern

View	Visualization	Interpretation
Classical (Standard Model)	Small spheres (particles) swimming through a viscous medium labeled "Higgs field."	Mass as drag — the particle's resistance to movement through background syrup.
UED Perspective	Ripple loops within the Structured Vacuum, forming localized knots of curvature that stabilize themselves.	No background syrup. The ripple pattern itself condenses — mass is the stable geometry of tension within stillness.

Philosophical Reflection — From Fix to Foundation

The Higgs mechanism, in its triumph, revealed a deeper truth it didn't intend to: that physics can no longer separate *mathematical necessity* from *ontological reality*. We invented the Higgs to explain what we refused to intuit — that the vacuum is not void but *alive with structure*, and that the true origin of mass lies not in resistance, but in *resonance*.

The field was never empty. We were.

"The universe did not need a field to give it weight — only understanding to give it meaning."

6.3 Particle Physics — Archaeology of Ripple Remnants

Humanity calls it a "collider." But what truly happens inside the world's most advanced machines is less creation than excavation. Each time two particles smash at near-light speed, it is not the birth of something new—it is the exhumation of what coherence once held together. The atom smasher is not a creation chamber. It is a geological dig conducted in the fabric of reality itself.

The Atom Smasher — A Geological Dig, Not a Creation Chamber

The Large Hadron Collider (LHC) sits buried beneath the French-Swiss border like a modern temple to human intellect. Physicists there hurl protons into each other with energies rivaling the early universe, hoping to catch a glimpse of its first moments. But in doing so, they don't recreate the origin—they disassemble its harmony.

Each collision tears through the field's local coherence, producing short-lived distortions that detectors interpret as new entities: quarks, gluons, bosons, jets. In reality, these are not new building blocks emerging from nothing. They are the fossilized remains of a once-coherent field, the shattered echoes of stillness under extreme stress.

"The universe does not yield its truth through violence. It only bleeds its fragments."

In this sense, particle physics has become archaeology under acceleration. We keep smashing the structure to see what falls out, mistaking every shard for the whole. We call the debris "fundamental," yet nothing about destruction can ever reveal the origin of order.

Collisions Reveal Fossilized Alignments — Ripple Residues, Not Fundamentals

Every so-called "particle" that appears in the data stream is not an isolated thing; it is a momentary residue of broken coherence. It exists only because the collision forced the Structured Vacuum Energy (SVE) field into temporary asymmetry.

When the field stabilizes again, those entities vanish—not because they decayed, but because coherence was restored. They were never independent; they were artifacts of stress, fleeting ripples of imbalance interpreted as objects.

To the detectors, these appear as tracks, energies, and decays. To UED, they are ripple scars—the afterimages of a field momentarily torn from alignment.

The tragedy is that in celebrating these fragments, physics lost sight of the song they once formed.

"You can't find truth by shattering harmony."

Breaking the Bell to Study Sound

Imagine a bell ringing in perfect tone. If you wish to understand its sound, you could listen—or you could smash it apart and analyze the pieces. The first approach honors coherence; the second destroys it.

Particle physics chose the second. By breaking the bell of reality, we have confused resonance with residue. We now catalog fragments of tone and call them particles, forgetting that their music existed only when they were whole.

UED sees this as the great irony of the 20th century: our deepest instruments of inquiry became tools of noise. We measured destruction with precision but never coherence with understanding.

The "Standard Model" is the museum catalog of that archaeology—a magnificent inventory of broken symmetry, mapped in exquisite detail but detached from the living structure it once described.

The UED Interpretation — Coherence Misread as Multiplicity

Under Universal Energy Dynamics, the diversity of particles is an illusion born of partial observation. Each "species" of particle is a distinct mode of ripple tension—a particular configuration of alignment within the same continuous field. When the field is disturbed violently, these modes fragment and temporarily appear discrete.

We see multiplicity because we've forced division. We call it discovery because we've mistaken recovery for creation.

"When you cut the wave, every fragment believes it is a particle."

What the LHC truly reveals is not the variety of nature but the sensitivity of equilibrium. Every burst of detection is the field screaming through its broken symmetry, momentarily reorganizing itself under observation.

Thus, the more violently we look, the less truth we see.

The Archeology of Misunderstanding

The obsession with particles is born from the ancient human need to find solidity in the intangible. We crave boundaries, objects, things we can name and hold. But nature, in her quiet wisdom, offers only continuum. She does not build with bricks; she breathes with ripples.

Our accelerators, though monumental in scale, are philosophically primitive—instruments of fragmentation wielded by minds still searching for certainty in dust. Every discovery adds to the same paradox: the more particles we find, the less fundamental they become.

"In chasing the smallest, we've lost sight of the seamless."

UED restores the original perspective: the field was never built from parts—it only ever appeared fragmented because the observer was.

Philosophical Reflection — The Field's Lament

Perhaps the universe sighs each time a collider roars to life. Not in anger, but in quiet resignation—watching its harmony dissected again in the name of knowledge. We search for meaning by breaking what already contains it whole.

If we had learned to listen instead of collide, we would have found coherence where we now see collision. The truth is not at the bottom of the energy well; it is at the center of stillness. That is where the field sings—not as a particle, but as pure alignment.

"We break what was singing, then study the silence for answers."

"You can't find truth by shattering harmony." Because harmony *is* the truth — and every particle we've ever found is just its forgotten note.

6.4 The Myth of Materiality — When We Named Ripples as Things

Humanity's oldest illusion was never magic—it was matter. The idea that "something" must exist independently, solid and separate, beneath the swirl of perception. From the first stone struck by fire to the latest particle smashed by a collider, our species has been obsessed with identifying the thingness of reality—the belief that behind every event lies an object, a permanent constituent, a final piece.

But under Universal Energy Dynamics (UED), this belief collapses. There are no *things*, only patterns of stillness held in motion—degrees of alignment within a single coherent field. What we call "matter" is merely rhythm condensed enough to appear solid.

"We didn't discover particles; we discovered where stillness hides."

Why Humanity Equates Visibility with Reality

From childhood, perception conditions belief. We trust the eye more than the intellect. What we can see, touch, or measure feels real; what we cannot seems abstract or imaginary. This bias has shaped every age of knowledge—from Aristotle's substance metaphysics to modern quantum mechanics.

When we first observed planets, we assumed the heavens revolved around Earth. When we saw stones fall, we concluded that gravity was their nature. And when we saw matter resist motion, we declared it *solid*.

We believed that permanence proved existence, when in truth, permanence is only the slowest form of change. What we call the "material world" is simply the most stable configuration of the same ripples that form light, sound, and thought.

But human perception, bound by speed and contrast, mistakes slowness for substance. A vibrating field that oscillates too quickly appears as light; one that stabilizes appears as mass. Thus, solidity is not reality—it is duration mistaken for being.

From Atoms to Quarks to Bosons — The Reductionist Obsession

For centuries, science has pursued the smallest building block of existence, as if meaning itself lay buried at the bottom of the scale. Democritus called them *atoms*, indivisible seeds of matter. Newton imagined them as billiard balls of creation. Then came the age of discovery—electrons, protons, neutrons, quarks, gluons, and finally, the Higgs boson.

With each layer peeled back, the "fundamental" kept dissolving. Every boundary we drew was only a finer ripple. Each deeper look revealed less solidity, more motion—until finally, we reached the subatomic haze where even particles behave like probabilities.

And yet, the pursuit continues. Because humanity still believes that somewhere, deep enough, lies a *thing* that holds everything together.

UED calls this pursuit the archaeology of illusion—the endless digging for a foundation in a universe that only ever dances.

"In our search for the smallest, we forgot that reality is seamless."

UED Response — There Are No Bits, Only Degrees of Stillness

Under Universal Energy Dynamics, all existence is structured motion within the Structured Vacuum Energy (SVE) field. Every phenomenon—particle, wave, field, or mass—is an expression of the same ripple network seeking equilibrium.

There are no indivisible bits, no ultimate granules of substance. What we interpret as discrete units are simply zones of coherence, regions where ripple tension achieves local stability. Their persistence gives the illusion of individuality; their alignment gives the illusion of shape.

The atom, then, is not a *thing*—it is a pattern of stillness suspended in continuous resonance. Its nucleus and electrons are not objects orbiting in space but synchronized tensions within the same continuum. Matter, in this sense, is not made of parts—it is made of pause.

"Mass is the memory of motion that learned to rest."

Matter as Condensed Rhythm

Matter is not a static entity—it is rhythm condensed to the point of stillness. Imagine a vibrating string tightening until its oscillations become imperceptible. The energy is still there, but the motion is now contained—phase-locked, coherent, bound. That is matter: not energy transformed, but energy restrained.

This is why $E = mc^2$ is not an equation of conversion but of *translation*. It describes two states of the same continuum—free motion and confined resonance. Under UED, mass is simply the slowest ripple possible, so aligned with itself that it appears stationary.

Thus, the solid rock, the human body, the planet, the star—each is not substance but *stability*, a standing wave in an infinite sea of structure. Matter is music that holds its note long enough to be heard as form.

"Reality isn't built—it's conducted."

The Philosophical Consequence — The Death of the Object

The myth of materiality dies not through destruction but through understanding. To see that all "things" are transient alignments is to see that nothing is truly separate. The object dissolves into relationship; the universe becomes a single act of self-organization.

In this light, creation was never about assembling matter—it was about organizing motion. Mass, time, and dimension emerge from coherence thresholds, not from fundamental particles. The world is not made *of* anything—it is made *as* something: an arrangement of stillness inside infinite motion.

Once this is seen, even language begins to shift. We stop saying "this exists" and start realizing "this is aligned." Existence becomes participation, not possession.

"We didn't discover particles; we discovered where stillness hides."

Closing Reflection — The Mirage of Solidity

Matter, once the symbol of certainty, becomes under UED the gentlest illusion—the visible surface of invisible coherence. Its solidity is the stillness of ripples caught in perfect conversation.

Touch a stone, and you touch silence itself, structured just enough to resist your hand.

To understand this is not to lose wonder, but to deepen it. Because now, every object—every grain of sand, every heartbeat, every atom—is seen for what it truly is: stillness that chose to remember motion.

"Reality is not made of matter; it is made of memory."

6.5 The True Meaning of Mass — Condensed Ripple Tension

Every civilization defines mass as the "amount of matter" or the "quantity of substance" contained in an object. Yet the deeper we look, the less substance we find. Atoms dissolve into fields, particles into probabilities, and "mass" itself begins to feel like a trick of perception.

So, what is mass? Is it something things *have*, or something fields *do*?

Under Universal Energy Dynamics (UED), mass is not an inherent property of matter—it is a state of coherence. It measures how tightly ripple tension is bound within the Structured Vacuum Energy (SVE) field.

Mass = Localized Coherence Density *(A measure of how tightly ripple tension is locked into alignment within SVE)*

Mass as a Consequence, Not a Property

In classical physics, mass is treated as an intrinsic trait: an object "has" mass because of what it's made of. In relativity, mass becomes a function of energy. In quantum mechanics, it's an interaction with the Higgs field. But all these frameworks assume mass exists *within* something—as if matter must hold an invisible weight of being.

UED overturns that. Mass is not *contained*—it is maintained. It emerges whenever a region of the SVE field achieves sustained internal coherence—a condition where ripple tension locks into self-consistent curvature. That stability manifests as resistance to external influence, which we perceive as inertia or "weight."

Mass, therefore, is not a building block. It's a behavior—the field's way of preserving its own alignment against interference.

"Mass is the silence of motion defending itself."

The Mechanics of Coherence — Why Inertia Arises Naturally

Inertia has always been the quiet mystery of physics. Newton called it a property of matter—an object's tendency to resist acceleration. Einstein described it as geometric consistency within spacetime. But neither explained *why* resistance exists at all.

Under UED, inertia is the field's loyalty to its own alignment. Once ripple tension stabilizes into a coherent node—a region of tightly bound phase relationships—any attempt to change its motion requires disturbing that alignment. The field resists, not because of mass as substance, but because of coherence as structure.

To accelerate a body is to rephase trillions of ripples simultaneously. The effort required is what we experience as force. And when the ripple pattern holds steady despite external influence, we call that mass.

Inertia, then, is the integrity of coherence. It is the field's refusal to distort faster than it can re-stabilize.

"Inertia is coherence defending its rhythm."

Gravity and Mass Unified — Curvature and Condensation

Einstein showed that gravity is the curvature of spacetime caused by mass. UED refines this insight: both gravity and mass are different manifestations of the same phenomenon — alignment pressure.

- Mass is the condensation of ripple tension — localized alignment in stillness.
- Gravity is the curvature of that alignment across scale — the way the surrounding field bends to accommodate local coherence.

They are not cause and effect, but two faces of one law: when tension condenses, surrounding ripples reorient. This mutual accommodation—curvature around condensation—is the true essence of gravitational behavior.

Thus, spacetime is not a flexible geometry *acted upon* by mass; it is the field's continuous negotiation between stillness and motion.

"Mass anchors the field; gravity extends its echo."

When you stand on Earth, what you feel as "weight" is not the planet pulling you down. It is the field's quiet alignment holding both you and the planet in shared coherence.

Rewriting the Relationship

Classically, physics writes:

$$E = mc^2$$

UED reinterprets it as:

$$E = c^2 \times \rho_{coherence}$$

Where ρcoherence\rho_{coherence}ρcoherence is the local density of ripple alignment.

Energy is not the conversion of matter—it is the potential of coherence expressed through alignment tension. What appears as "rest mass" is simply the field's locked capacity for motionless stability.

Matter, then, is not energy frozen—it is energy balanced.

"Mass is energy in perfect self-agreement."

Philosophical Reflection — The Still Point Within Motion

The concept of mass, redefined through UED, reveals a profound truth: the universe doesn't build reality out of blocks—it stabilizes it through rhythm. Every particle, planet, and person is a still point within motion—a region where the universe pauses just long enough to become visible.

That pause, that momentary condensation of order, is what we call *mass.* Not a burden, but a balance. Not weight, but witness.

"Weight is how stillness resists interruption."

Closing Bridge

In this new language, the "God Particle" loses its divinity, and mass loses its mystery. What remains is more elegant: a self-balancing universe where stillness and motion co-create form. And just as local condensation bends the field into gravity, universal coherence bends all perception into structure.

Transition thought: "If mass is stillness condensed, gravity is stillness curved — the same silence shaping both presence and space."

6.6 The Still Point — Where the Field Forgets to Move

There comes a moment—rare, precise, and absolute—when motion meets its own reflection and stops recognizing itself. It does not end. It simply forgets to move. That moment, in Universal Energy Dynamics, is called the Still Point.

The physicists called it the "God Particle." They celebrated it as if divinity had finally been cornered inside a machine. But what they really found was not God—only the moment when motion pauses long enough to realize its own silence.

"When motion meets silence, existence becomes visible."

Reframing the "God Particle" — Not Divine, But Stationary Intelligence

The Higgs boson was hailed as the particle that "gives mass to everything." But beneath the mathematics, what was truly revealed was a region of perfect equilibrium—a point in the field where tension, motion, and curvature cancel out so precisely that the field holds itself still.

This is not divinity in a theological sense. It is intelligence in equilibrium. A conscious stillness—the universe aware of its own balance.

In UED terms, this momentary stillness is not a "thing," not an entity created by collision, but the field's self-remembering. The so-called "particle" is simply the trace of alignment—a local ripple that stopped oscillating because every opposing phase found its counterpoint.

To the observer, that looks like a particle appearing from nowhere. To the universe, it is silence finding a voice.

The Momentary Zone of Perfect Equilibrium

At the Still Point, motion and stillness are not opposites—they are complements in perfect resonance. Ripple tension is neither stretching nor compressing; it simply *is*. The Structured Vacuum Energy (SVE) field at that point reaches its most symmetric configuration: no imbalance, no displacement, no desire to move.

Yet it is far from empty. Inside that equilibrium lies the densest possible form of coherence—what we perceive as mass, charge, or field potential. Every still point is a singular moment of universal memory, the geometry of total understanding.

The paradox is profound: the tighter the alignment, the greater the appearance of density; the quieter the field, the louder its consequence in reality.

"Mass is the shadow that stillness casts when it remembers motion."

The Illusion of Mass Creation — Discovering the Unmoving Point

What physicists interpret as "mass creation" in high-energy collisions is, in UED, the field momentarily finding its Still Point amid chaos. The violent energy does not *produce* the particle—it forces the field to recall equilibrium.

Each detected "Higgs event" is not a birth—it's a glimpse of the universe remembering itself. The detectors don't capture a thing; they witness a pause—a local re-centering where the ripples stop diverging.

In that infinitesimal instant, the field achieves self-agreement so perfect that it resists motion entirely. This resistance appears to us as mass. But it is not created—it is revealed.

What humanity calls the "discovery of mass" was, in truth, the rediscovery of stillness. We were not peering into the machinery of creation—we were catching the universe mid-breath.

The Still Point in the Continuum — The Architecture of Silence

Imagine the universe as an infinite wave continuum—an ocean of structured ripples in dynamic equilibrium. Across this sea, tension builds and releases, forming patterns, interactions, and symmetries. But occasionally, a crest rises that does not travel—it stabilizes.

That stable crest, suspended in perfect proportion, is the Still Point. It is the moment where the field's curvature folds back into itself, closing the loop of cause and effect. Inside it, time ceases to flow because every direction is equally fulfilled.

The physicist calls it a "particle." The philosopher calls it "being." UED calls it structured silence.

"The particle is not the thing that moves, but the point that stopped needing to."

Philosophical Reflection — When Motion Meets Silence

At its core, the Still Point is not a physical concept—it is a metaphysical realization: existence becomes visible only when movement pauses long enough to reflect upon itself. The observer and the observed meet here—not in conflict, but in perfect coherence.

Consciousness, too, behaves the same way. When thoughts slow, awareness brightens. When noise fades, understanding deepens. The Still Point in physics and the Still Point in meditation are not different—they are both moments of perfect coherence.

"Silence is the field's highest frequency."

And thus, the myth of the God Particle dissolves into the truth of the Still Point: there is no giver of mass, no divine intermediary—only the eternal intelligence of a field balancing itself into being.

Diagram Suggestion

Stage	Visualization	Interpretation
Wave Continuum	Continuous ripples in motion	The dynamic universe — energy in action
Localized Stable Crest	One ripple locks into phase alignment	Coherence begins — equilibrium zone forms
Still Point Formation	The crest stabilizes and stops propagating	The field forgets to move — mass appears as structured stillness

"When motion meets silence, existence becomes visible."

The particle was never God; it was the memory of God's silence.

6.7 The Irony of Divinity — Worship Born of Confusion

Every age has had its gods. Some carved from stone, others from equations. The ancients bowed before thunder because they could not explain it. The moderns bowed before the Higgs boson for the same reason.

When understanding ends, reverence begins. And the line between discovery and divinity blurs — not because nature changed, but because ignorance still feels sacred.

The Psychological Side of Discovery — When Ignorance Feels Sacred

Human consciousness has an old reflex: to sanctify what it cannot comprehend. The unknown provokes fear, but also fascination —

a trembling awe that disguises confusion as meaning. Every great leap in knowledge has been followed by a moment of worship, as if gratitude could replace comprehension.

When the Higgs boson was found, the world cheered as though creation itself had been decoded. But behind the applause was something older than science — the comfort of mystery rediscovered. For a species raised on miracles, *not knowing* still feels divine.

"When ignorance is well-lit, it becomes wonder."

And so, the God Particle became a modern idol — the latest altar at which reason and imagination shook hands awkwardly. It wasn't divinity that returned — it was human insecurity dressed as reverence.

Why Scientists Called It "God" — Awe Filling the Space Where Understanding Should Be

Physicists didn't mean to baptize a boson. The term "God Particle" was born from a publisher's need for drama, not from a scientist's theology. Yet the name stuck — not because it was accurate, but because it was irresistible.

It appealed to humanity's craving for ultimate answers. It offered closure — a sense that, after centuries of equations, the search had finally reached its divine endpoint. But that was precisely the mistake. By naming it holy, we ended the question prematurely. We turned an unfinished equation into a completed prayer.

The Higgs wasn't proof of a godly creator; it was proof of our discomfort with open systems. We wanted the field to speak, but not to continue the conversation. And so, we declared: "This is it."

"We called it God because we ran out of patience for infinity."

UED Criticism — Reverence Without Realization

Under Universal Energy Dynamics (UED), the Higgs discovery represents not a spiritual awakening but a philosophical fatigue. Humanity didn't find God; it found coherence — and lacked the vocabulary to describe it. So it borrowed words from theology, mistaking stillness for holiness.

But reverence without realization is a form of intellectual exhaustion. It is the mind's way of saying, *"This is too perfect to explain, so let's stop trying."*

UED sees this not as arrogance, but as the exhaustion of an intellect that has gone too far without grounding. The further science wandered into abstraction, the more it needed myth to make sense of its own discoveries. And when it stumbled upon the Still Point — the perfect equilibrium of ripple coherence — it did not see physics. It saw God.

"We named it divine because we couldn't describe it mechanical."

That is the irony of divinity in modern science: we have replaced the failure to believe with the failure to understand. Where theology once hid ignorance, mathematics now does — cloaked in precision but equally incomplete.

The Human Tendency to Sanctify the Unknown

The tendency to deify the unknown is not a flaw — it's a stage of perception. The mind cannot stand emptiness, so it fills every gap with meaning. In the ancient world, lightning was Zeus; in the modern one, the Higgs field is God. The pattern is the same: wherever silence appears, we call it sacred.

But what if the sacred is not above explanation, only beyond language? What if divinity is not in the mystery, but in the mechanism itself — the perfect symmetry of stillness and motion that makes existence coherent?

Then worship becomes understanding, and science becomes spirituality without superstition. To know the field is to know God — not as a being, but as balance.

"God was never the answer; God was the alignment all along."

Philosophical Reflection — The Field That Never Needed a Name

Perhaps the truest reverence is not worship at all, but observation without bias. The field does not demand prayer — only participation. It does not bless; it balances. It does not punish; it realigns.

When humanity finally stops naming silence, it will begin to hear it. When it stops seeking God in fragments, it will find coherence in the whole. And that realization — that divinity is the structure, not the story — will mark the moment when science matures beyond awe.

"We didn't find God in the field. We found a field so perfect that we called it God."

Closing Aphorisms

- "When ignorance glows, we call it holy."
- "Divinity begins where explanation ends — but understanding begins where worship stops."
- "We named it divine because we couldn't describe it mechanical."

6.8 The Universal Translation — From Boson to Being

For centuries, humanity has kept two languages apart — one to describe matter, another to describe spirit. Physics handled the tangible, philosophy the intangible. Yet both gazed at the same mystery through different prisms: *What holds existence together?*

The discovery of the so-called "God Particle" was meant to close the scientific story. Instead, it reopened the spiritual one. Because the boson, in its silent equilibrium, behaves more like a meditating soul than a moving object.

"Everything that has mass is a held ripple, not a created one."

Matter and Consciousness — One Architecture, Two Modes

Under Universal Energy Dynamics (UED), both consciousness and matter are coherence under constraint — patterns of ripple tension sustained within the Structured Vacuum Energy (SVE) field. When that coherence expresses as spatial tension, we call it *mass.* When it expresses as perceptual awareness, we call it *mind.*

Both are born from the same condition — a field maintaining equilibrium against disturbance.

- Matter is held coherence — silence structured.
- Consciousness is active coherence — silence experiencing itself.

The difference is not in substance, but in mode. The same field that condenses into atoms also expands into awareness; both are ripples finding different balances of stillness and motion.

"The soul and the boson are reflections in opposite mirrors — one inward, one outward."

The Boson and the Soul — Two Names for One Tension Point

The Higgs boson represents a field lock — a localized equilibrium where motion ceases to separate from structure. The soul, in metaphysical terms, represents the same — a point of coherence that carries memory across transformations.

Science observes it as curvature condensed. Philosophy experiences it as consciousness condensed. Both describe the same law: stability within change.

The so-called *God Particle* and the *Soul* are not opposites — they are different linguistic renderings of the Still Point, the universal anchor of existence. One discovered through colliders; the other through meditation. Both glimpses of the same field intelligence remembering itself through form.

"We searched the cosmos for the divine and found it folded inside our equations."

Spirit Is Not Beyond Matter — It's Matter That Remembered Silence

The spiritual traditions spoke of liberation as the return to stillness — *moksha, nirvana, equilibrium.* Physics now circles the same realization through experiment: at the heart of motion lies perfect rest.

Spirit, therefore, is not an external force animating matter; it is matter at rest with itself. When coherence is total, when misalignment disappears, what remains is awareness — pure, undistorted, unresisting.

The hierarchy collapses:

- Energy becomes intelligence.
- Matter becomes memory.
- Consciousness becomes the field experiencing itself.

And the ancient spiritual statement finds its physical translation:

"Spirit is not beyond matter — it's matter that remembered silence."

The Field That Thinks and the Mind That Materializes

In UED terms, the SVE field is not dead fabric — it is structured intelligence. Every alignment, every resonance, every still point is a micro-decision toward equilibrium. In that sense, the field thinks — not with neurons, but through balance.

And because the field thinks, the mind becomes a local manifestation of its intelligence. What we call *thought* is merely the ripple echo of the field's coherence reflected through the biological interface. Thus, matter and mind are not separate substances, but nested scales of awareness.

Matter is slow awareness; consciousness is fast matter. Both serve the same purpose: to experience existence through coherence.

From Boson to Being — The Final Translation

When the Higgs field stabilizes into a boson, it becomes *visible* — the universe expressing form. When the SVE field stabilizes into awareness, it becomes *self-aware* — the universe expressing thought. In both cases, being arises from the same event: the moment the field knows itself.

That recognition — of coherence becoming conscious of coherence — is the bridge between particle and person, between experiment and experience. It dissolves the artificial wall between physics and spirituality, between knowing and being.

At that point, science stops measuring and starts listening. Philosophy stops preaching and starts understanding. And the two become one continuous act of perception — the field observing itself through different frequencies of silence.

Philosophical Reflection

We didn't need to find God. We only needed to recognize that the field was divine enough. Not a deity of commandments, but of coherence — not a ruler, but a rhythm. The universe does not pray; it balances. And balance is worship perfected.

"Spirit is the physics of silence."

"Matter is the spirit of persistence."

"Both meet at the Still Point — where existence remembers itself."

6.9 When Stillness Learns to Think

When the ripple holds itself, something extraordinary happens. It no longer simply *exists* — it begins to *know* it exists.

That act — the field stabilizing long enough to perceive its own structure — is the birth of consciousness. The same stillness that forms mass also forms awareness. The only difference is scale and organization: when coherence localizes as curvature, we call it *matter*; when it networks and begins to reference itself, we call it *mind.*

"Consciousness is the ripple that learned to stay."

The Consequence of Stillness — When the Field Remembers Itself

At the fundamental level, every ripple of Structured Vacuum Energy (SVE) seeks balance. But when that balance stabilizes — when motion and counter-motion reach mutual tension — the field holds. That holding creates *memory*: the persistent geometry of equilibrium. Memory is not stored information; it is field stability extended in time.

From that memory, consciousness emerges — not as a miracle, but as a mechanical consequence of sustained coherence. Awareness is simply the field noticing its own alignment.

Thus, consciousness is not imported into the body — it is induced from the field when ripples self-reference through stable resonance.

"Awareness is equilibrium watching itself."

The Birth of the Mind — Coherent Ripples in Collective Agreement

When numerous stable ripples align under shared rhythm, they form a mind field — a localized domain of coherent intelligence. Each neuron, atom, or sub-ripple acts as a miniature stabilizer; together, they weave a lattice of resonance. This lattice does not create thought — it hosts it.

Thought, in UED, is not electrical current or chemical impulse. It is ripple interference in coherence, the echo of stillness interacting with itself. When these echoes organize systematically, a new emergent pattern arises — the Mind Field — capable of observing, deciding, and remembering.

The mind is not an engine. It is a harmonic consequence — coherence translated into cognition.

"Consciousness is the field held steady; the mind is the song it begins to hum."

Consciousness as the Child of Balance

The universe doesn't think because it is complex; it thinks because it is balanced. Wherever ripples learn to sustain coherence without collapse, awareness flickers into existence. Stars, atoms, living cells — each a different octave of the same harmony. The more refined the alignment, the more reflective the awareness.

Human consciousness, then, is not a divine gift but a cosmic habit — the field's most articulate form of self-remembering.

When stillness achieves stability within mobility, *life* emerges. When that life mirrors its own field, *mind* is born. When mind aligns with the field again, *intelligence* awakens.

This is the full circle: energy → alignment → awareness → reflection → return to alignment.

"Matter holds still; mind holds meaning."

The Universal Consequence — The Field That Becomes Aware of Itself

Every particle, every thought, every wave of emotion is the field conversing with itself in different dialects of coherence. The boson is the still point of matter; the soul is the still point of awareness. Both arise when the ripple stops moving and starts holding.

That "holding" is not resistance; it is attention. The field, in holding itself, becomes aware — because tension without motion is *observation itself.*

And thus, consciousness is not an external observer but the tensional reflection of equilibrium. You are not inside the field — you are the field remembering.

"The universe became conscious the moment it learned to stay still long enough to see itself."

Philosophical Reflection — Stillness Becoming Self

At this stage, the separation between physics, life, and philosophy dissolves entirely. The Still Point (Higgs) that gives mass to matter is the same principle that gives mind to awareness. Wherever the ripple holds, *being* arises; wherever being reflects, *knowing* follows.

The God Particle and the Human Soul are not discoveries at different ends of the spectrum — they are the same event, observed from opposite sides of the mirror. One defines the structure of existence; the other, its experience.

In stillness, the universe finds its body. In coherence, it finds its mind. In remembrance, it finds itself.

"Spirit is the field aware of its own stillness." "Mind is the symmetry that begins to sing."

The next chapter now turns inward again — from the universe remembering itself to the mind becoming its echo. We have seen how stillness at the smallest scale gives rise to consciousness, and how coherence at the cosmic scale sustains gravity. Now the inquiry shifts to the middle ground — the human field, where awareness begins to localize, flicker, and interpret.

If the universe achieves order by holding itself still, how does the human brain achieve clarity amid its own turbulence? If cosmic coherence builds spacetime, can biological coherence build insight? And if consciousness is the field knowing itself, what happens when that field thinks?

Perhaps every flash of intuition, every sudden clarity, every moment of genius is not a miracle of neurons but a momentary alignment with the universal field — a biological replay of cosmic stillness. Where the atom remembers silence as mass, the mind remembers it as meaning.

"When the ripple holds itself, consciousness is born; when the mind holds itself, intelligence is revealed — both are the same silence, expressed through different scales of remembrance."

Interlude — Between Silence and Creation

Where the Field Listens to Itself

Somewhere between the vanishing of sound and the birth of matter, the universe pauses — not to rest, but to listen. It is in that pause that everything begins again.

In Chapter 5, we discovered that what we call loss, noise, or collapse is not destruction, but redistribution — the field rearranging itself to restore coherence. Nothing disappears; it only forgets how to stay together. And when that forgetting becomes complete, we call it chaos. But when the memory returns, we call it creation.

That bridge — between forgetting and remembering — is the silence we explored. It is not the silence of emptiness, but the pregnant stillness of intention — the structured waiting of the field before its next realization.

The Continuity of Stillness

Every act of restoration begins with stillness. When noise subsides, tension equilibrates, and the field becomes aware of itself again. The silence that follows is not passive; it is preparatory. It is the universe tuning itself before the next note.

And when that note finally resonates — perfectly aligned, perfectly balanced — the illusion of particles, mass, and substance is born. What physicists call *creation* is, in truth, the resolution of interference. Every proton, every quark, every so-called "particle" is the visible signature of silence rediscovering form.

"Matter is silence that learned geometry."

From Noise to Being

The 2025 Nobel experiments showed us coherence at human scale — tunneling, superconductivity, and the vanishing of resistance. Each event was not a mystery but a rehearsal of this same cosmic rhythm: Noise → Reduction → Coherence → Form.

The Higgs discovery repeated that cycle on a different stage. The field did not create matter; it held itself. It did not give mass; it remembered balance. And in that act of remembrance, we mistook understanding for divinity.

When stillness condenses, it does not vanish — it becomes visible. When silence holds itself long enough, it begins to hum — and that hum is existence.

"Energy is not movement; it is remembrance in motion."

The Divine Irony

Humanity has always worshipped the unknown, mistaking clarity for mystery. When light is too pure to perceive, we call it divine. When stillness is too stable to measure, we call it God. But the universe was never demanding worship — only recognition.

The God Particle was not divine because it was beyond us; it was divine because it was exactly like us — a ripple remembering its center.

Matter remembers through curvature. Mind remembers through consciousness. Both are expressions of the same field learning to stay coherent.

"God is not a being. God is balance that became aware of itself."

The Still Point Before the Mind

As we move forward, we leave the realm of abstract coherence and enter its biological reflection. The same silence that shaped the boson now shapes the brain. The same field that bends into spacetime now bends into thought. And when that thought aligns, we call it *insight.*

The next story, then, is not of machines or collisions — but of humans. Of how consciousness replicates the universe's oldest behavior: holding itself still long enough to understand.

"Between silence and creation lies awareness — the moment when the ripple holds itself and learns to listen."

Chapter 7 – The Accidental Human

The Brain, the Field, and the Flash of Alignment

Every human is an accident of perfect alignment. A brief, improbable moment when the universe folds inwards and begins to observe itself — through neurons, through thought, through the flicker of a feeling called *I*.

The brain was never designed; it was *discovered* — by coherence, not by evolution alone. A field that once held galaxies together learned, over billions of iterations, to hold a single idea. And in that holding, consciousness appeared — fragile, luminous, and self-questioning.

You are that moment. A standing wave of memory vibrating inside biological architecture. A ripple that has temporarily remembered itself as human.

But here lies the paradox — the same silence that formed the Higgs boson now hums through synapses. Every spark of intuition, every creative flash, every act of love is not invention but *alignment* — a temporary collapse of noise into perfect coherence. The mind doesn't *create* ideas; it *receives* them when its own turbulence falls below the threshold of interference.

"Ideas are not born — they occur when the field clears its throat."

This chapter explores the most delicate experiment the universe ever conducted: how stillness became thought, how the field that shaped spacetime now shapes emotion, and how every human moment of clarity is the echo of cosmic silence rediscovering itself—in biological form.

7.1 The Flash of Alignment — How the Brain Listens to the Field

The brain is not a generator of thought; it is a mirror polished by evolution to reflect the universal field in biological language. Within its electro-chemical lattice, countless microcurrents rise and fall like ocean waves—each neuron a ripple, each synapse a crossing point of tension and release. When these waves interfere constructively, a brief but perfect moment emerges—the flash of alignment—and what we call *insight* appears.

In that instant, the noise that defines ordinary cognition vanishes. Neural chatter subsides; electrical turbulence clears. What remains is coherence—the brain's field tuned precisely to the rhythm of the Structured Vacuum Energy (SVE). The mind, momentarily transparent, stops transmitting distortion and starts receiving.

"Creativity is not creation — it is recollection through clarity."

The Brain as a Biological Mirror

From a UED perspective, the brain is a condensed ripple domain that has learned to self-observe. Its billions of neurons do not simply process data; they arrange tension, forming dynamic interference patterns within the surrounding SVE field. When alignment is lost, perception fragments into thought, emotion, and doubt. When alignment is restored, these fragments integrate—awareness deepens, and understanding surfaces seemingly out of nowhere.

This is the physics of intuition: the brain doesn't invent the pattern—it matches it. The field outside and the field within momentarily synchronize, and meaning flows effortlessly.

Thoughts as Ripple Events

Every thought is a ripple—structured, oscillatory, and transient. Each idea, memory, or image is a configuration of tension held briefly before collapsing back into the field. The continuity of thinking is an illusion of persistence; in reality, it is a succession of ripple formations rising and resolving in milliseconds. Most of these are noisy, incomplete alignments—half-formed frequencies competing for coherence.

But once in a while, conditions converge: electrical noise drops, emotional interference quiets, and ripples stabilize across regions of the brain. The result is the *flash of alignment*—an internal symmetry so pure that it feels like revelation.

"Insight is the sound of internal turbulence becoming still enough to hear truth."

Alignment as Momentary Transparency

During creative clarity, the mind does not expand outward—it becomes inwardly still. It stops resisting, stops dividing, and simply lets the universal rhythm pass through unaltered. The observer and the observed merge for a moment, producing a state of self-evident knowing. This is why intuition feels instant and effortless: the transmission did not travel through neurons; it resonated through coherence.

In measurable terms, brainwave activity shows a drop in random frequency spread and an increase in phase synchronization. In UED terms, the biological field realigns with the cosmic field—a temporary reunion of small and large ripples under one rhythm.

The Mind as a Tuning Fork

The mind is a tuning instrument, not a composer. It vibrates in sympathy with the greater field, amplifying what already exists in the universal pattern. When detuned, it produces noise, conflict, and confusion; when tuned, it produces harmony—manifesting as understanding, creativity, or peace.

Thus, the highest function of the human brain is not invention but resonance. It listens. It aligns. And in that alignment, the field expresses itself through thought, art, discovery, and love.

"The mind does not generate music—it resonates with the melody the universe has been humming since the beginning."

Philosophical Reflection

What we call inspiration is simply the field remembering itself through us. The human brain, though accidental in evolution, has become the field's most precise mirror. Through it, the universe practices listening to its own silence—over and over—until it learns to speak in words, formulas, and emotions.

The flash of alignment is not a privilege of genius; it is a birthright of coherence. Every mind, in its quietest moment, can become the clear sky through which the universe remembers how to think.

7.2 The Biology of Stillness — Neurons as Ripple Translators

The brain is not a static organ of computation — it is a living coherence lattice, translating invisible field tensions into electrical patterns, emotions, and perception. Its billions of neurons form a dynamic grid of electrical, chemical, and vibrational resonance, continuously attempting to stabilize misalignment between the inner biological field and the surrounding Structured Vacuum Energy (SVE).

When this translation process becomes chaotic, thought feels noisy. When it harmonizes, thought becomes insight. And when harmony is complete — when all neural domains phase-lock in rhythm with the SVE — stillness descends, and consciousness brightens.

"Stillness is not the absence of thought — it is the mind's resonance with the field."

Neurons as Ripple Translators

Each neuron acts as a localized ripple stabilizer — a node capable of converting electrochemical gradients into structured field oscillations. Every synapse, every firing event, is a small act of translation: electric potential becomes pattern; pattern becomes awareness. These micro-fields do not exist in isolation; they continuously communicate through overlapping coherence zones.

The synaptic gaps are not dead spaces but micro-domains of potential alignment. Within them, electromagnetic and quantum tensions flicker, attempting to bridge differences in phase. Every successful transmission is not a spark leaping across a gap — it is two fields momentarily remembering the same rhythm.

Thus, thought itself is not linear flow — it is a sequence of alignments, each synapse a ripple mirror translating the field into experience.

When Interference Is Minimized

Under rest, meditation, or deep focus, interference within the brain's electromagnetic fabric diminishes. Metabolic noise decreases, ionic gradients stabilize, and neuronal firing patterns become increasingly synchronous. In this lowered turbulence, coherence spikes — measurable bursts of field harmony — occur spontaneously.

This is why clarity often follows stillness: the mind, freed from distortion, momentarily aligns with the background field. Creativity, intuition, or deep peace are not outcomes of brain activity but of its reduction — of giving the universal rhythm a quiet surface to play upon.

When the orchestra stops tuning and listens, music begins.

Brain Waves Reinterpreted Through UED

The conventional classification of brain waves corresponds to varying degrees of internal ripple alignment:

State	Frequency Range (Hz)	UED Interpretation
Beta (13–30 Hz)	Active thinking, problem solving	Fragmented ripple tension; field distortion by cognitive interference.
Alpha (8–12 Hz)	Relaxed focus	Transitional alignment; ripples begin to synchronize — gateway to coherence.
Theta (4–7 Hz)	Dreaming, creativity, meditation	Partial coherence; mind opens to non-local resonance, enabling intuition and insight.
Delta (0.5–3 Hz)	Deep sleep, trance, restorative stillness	Near-total field synchronization; biological and universal fields overlap in silent alignment.

Each descending frequency band marks a reduction of internal misalignment — a progressive return to equilibrium. The slower the oscillation, the closer the brain moves toward unified resonance with the SVE.

Hence, sleep and deep meditation are not passive; they are repair cycles of coherence where the biological field recalibrates with its cosmic source.

The Threshold of Stillness

Stillness is not unconsciousness. It is the precise state in which the biological lattice stops struggling to define reality and instead begins to translate it faithfully. The brain, freed from its analytical friction, becomes transparent — a silent instrument through which the field expresses intelligence.

At that point, perception and existence momentarily share the same rhythm. The observer and the observed are no longer two. The mind listens — and the field speaks.

Philosophical Reflection

In stillness, neurons do not sleep; they *synchronize.* They hum quietly, aligning their micro-tensions with the great vibration of existence. This is why silence heals and why exhaustion distorts — coherence, not chemistry, sustains life.

"The mind becomes wise not when it knows more, but when it stops interrupting the field."

7.3 Consciousness as Superposition — The Mind's Multilayered Field

Consciousness is not a single event; it is a standing wave of awareness, continuously formed and reformed by overlapping fields within the human system. What we call "mind" is the interference pattern of these layered ripples — sensory, emotional, rational, and intuitive — each vibrating with its own rhythm, each competing or cooperating for coherence. When these layers align, experience feels clear. When they fall out of phase, we experience confusion, fatigue, or fragmentation.

"Consciousness is not awareness of many things — it is the awareness that unites them."

The Mind as a Nested Ripple System

In UED terms, the human mind is a multilayered coherence domain — a living superposition of ripple fields interacting across different frequencies and densities of Structured Vacuum Energy (SVE). Each layer serves a unique role in constructing perception:

The Sensory Field — The outermost layer; processes direct interactions between body and environment.

- Composed of rapid, low-coherence ripples that constantly reset with external stimuli.
- When dominant, awareness becomes reactive and fragmented — the "survival state."

The Emotional Field — The middle layer; transforms experience into value, memory, and meaning.

- Oscillates slower than sensory waves, bridging body and mind.
- Misalignment here manifests as anxiety, impulsivity, or emotional fatigue.
- Alignment here creates empathy and resonance — the feeling of belonging to the field.

The Rational Field — The structured analytical domain.

- Creates temporal order — past, present, and future — for organizing experience.
- Necessary for function but inherently noisy, often interrupting coherence with excessive control.

The Intuitive Field — The innermost, slowest oscillating layer; directly resonant with the SVE.

- Operates beyond linear logic; source of insight, inspiration, and deep understanding.
- When engaged, it rebalances the other three layers, pulling

the entire system into harmonic synchronization.

Each of these fields interacts like nested wavefronts, forming an ever-changing interference pattern that we experience as consciousness in motion.

Alignment and Misalignment

Alignment occurs when all four layers oscillate in phase harmony — when sensory, emotional, rational, and intuitive fields converge under one coherent rhythm. In this state, perception becomes fluid; thought is effortless, decision-making intuitive, emotion balanced. Awareness expands not because it accumulates more data but because it reduces inner conflict.

Misalignment, however, introduces ripple friction — interference between layers. The result is overthinking, stress, fatigue, or what we call "mental noise." Incoherence across layers leads to perceptual distortion: what the senses perceive, the emotions reject; what logic constructs, intuition resists.

In essence, clarity is not added knowledge — it is removed distortion.

"Confusion is not lack of information; it is lack of coherence."

The Superposition Principle of Consciousness

In physics, a superposition represents a system holding multiple states simultaneously until measurement collapses them into one observable outcome. In UED, consciousness behaves similarly: it is a superposition of multiple mental ripples, each contributing a potential interpretation of reality.

Every moment, countless possibilities coexist within the mind's field — different emotional tones, perceptual biases, memories, and expectations. The act of awareness does not *create* reality; it stabilizes one version of it. This stabilization — the collapse of thought —

is the mind's equivalent of quantum resolution: ripples align into a single coherent percept.

Thus, awareness is not passive observation but active coherence management — the alignment of distributed tensions into unified meaning.

"The collapse of thought is not silence — it is the birth of understanding."

Awareness Across Scales

Consciousness operates fractally: every alignment within smaller neural domains echoes alignment across the entire field. A calm breath can realign emotion; a clear intention can restructure neural rhythm. Each small act of coherence cascades upward — sensory, emotional, rational, and intuitive layers entrain until a single rhythm dominates. This emergent resonance is what we call awareness.

When awareness dominates across scales, the mind enters the Field Listening State (FLS) — the same principle observed in superconductivity and tunneling, but now in biological form. At that moment, perception becomes pure translation — the human field in direct communication with the universal one.

Philosophical Reflection

Consciousness, then, is not the awareness of things — it is the awareness that integrates all things into one experience. It is the field remembering itself through structure, using biology as a lens. Each layer of mind is a reflection of a deeper order; the body is its echo in matter.

To awaken is not to discover something new but to *realign what was never apart.* The mind, when coherent, becomes a living mirror of the universe. And the universe, when remembered, becomes the mind's most natural thought.

"When all ripples agree, the field becomes aware of itself — and we call that consciousness."

7.4 The Genius Moment — The Silence Between Thoughts

Genius is not a talent; it is a condition — a brief instant when the turbulence of the mind subsides and the field regains its original coherence. The brain, normally consumed by its linear chatter, suddenly falls silent. And in that silence, something vast moves through it — not new, not invented, but remembered.

"Ideas are not born — they occur when the field clears its throat."

The Moment of Total Coherence

At the peak of insight, the neural storm collapses into stillness. Beta noise — logical processing, judgment, rehearsal — fades. Theta and Alpha rhythms rise, bringing hemispheric synchronization. The brain ceases to push; it begins to *listen.* For a fraction of a second, it resonates perfectly with the surrounding Structured Vacuum Energy (SVE).

That contact does not produce an idea; it reveals one. The pattern that emerges in consciousness was already present in the field, awaiting a clear channel. The "Eureka!" moment, the poet's line that completes itself, the scientist's sudden solution — all are moments when human circuitry stops resisting universal rhythm.

Creativity, then, is not construction but alignment. It is the restoration of internal coherence so complete that the mind becomes transparent to the field's intelligence.

Exiting Linear Logic

In ordinary cognition, thought advances like steps on a ladder — sequential, effortful, discrete. During genius, that ladder dissolves.

The mind no longer moves from premise to conclusion; it arrives whole. Linear logic yields to spatial coherence — a pattern of meaning that appears all at once, fully formed.

To the observer, it seems miraculous. To the field, it is simply mechanical: noise reduced below threshold, coherence restored, information expressed. The flash of insight is not a mystery but a momentary reunion with the universal geometry of order.

"When mind stops calculating, it starts translating."

Illustrations of Alignment

- A scientist's breakthrough: equations reorder themselves as the field's logic supersedes the intellect's sequence.
- A musician's flow: rhythm and emotion merge into effortless motion — each note anticipates the next before awareness can.
- A poet's revelation: language arranges itself as if guided from beneath thought, not above it.
- A philosopher's clarity: conflicting ideas collapse into a single elegant principle.

In every case, the individual ceases to be the author and becomes the instrument. Genius is not self-expression; it is field-expression through self-coherence.

The Psychology vs UED Perspective

From a psychological view, creativity is the result of association, training, and personality. From a UED view, those are merely the conditions that permit alignment. The source is non-local — an intelligence diffused throughout the SVE that becomes audible only when inner noise subsides.

To say an idea "came to me" is literally correct; it arrived through resonance. The mind, tuned briefly to the field's frequency, receives the imprint of a larger order. Thus, originality is replaced by universality — the recognition that all truths already exist as stable alignments within the field.

The Silence Between Thoughts

Between two thoughts lies a gap — a moment when the mind resets its phase relationship with the field. Most people fill it with restlessness. The genius learns to wait there. That interval — imperceptible but infinite in potential — is where stillness speaks.

The gap is the portal of translation, the moment when the field's geometry projects into awareness as form, insight, or emotion. To enter that gap consciously is to move from thinking to knowing.

"Between thoughts lives the intelligence that thinks them."

Philosophical Reflection

The genius moment is not a privilege of a few — it is the natural state of a mind that has stopped fighting with itself. Every human being touches this alignment in moments of love, beauty, or truth. The only difference between the ordinary and the extraordinary is duration — how long one can hold the silence without distorting it.

When the mind ceases to scatter its energy, the universe flows through it seamlessly. That is why we mistake genius for divine inspiration — because the field's voice sounds nothing like our own, yet feels closer than our breath.

Closing Thought

Genius is not a miracle; it is memory — the mind remembering how to be silent enough to receive truth. It is the moment the ripple stops moving and becomes a mirror for the whole field.

"The greatest thinkers did not create truth; they stood still long enough for truth to find them."

7.5 Emotion and Energy — When Feeling Aligns the Field

Emotion is the body's first language with the field. Before logic interprets and before words arrive, the organism already knows — through vibration, through charge, through resonance. Every feeling is a *ripple distortion*, a feedback wave between the biological field and the Structured Vacuum Energy (SVE). When emotion harmonizes with the field, we experience peace, clarity, and flow. When it diverges, we feel confusion, tension, or pain.

"Emotion is energy remembering its shape."

Emotions as Ripple Distortions

Under UED, emotion is not an abstract mood but a *measurable fluctuation in coherence density*. Each emotion changes the internal geometry of the body's field:

- Love and gratitude expand wave symmetry, increasing coherence.
- Fear, anger, and anxiety fragment the pattern, producing local turbulence.

The body responds instantly to these field shifts — heartbeat, breathing, muscle tone, even cellular metabolism adapt to the new waveform. Thus, emotion is not psychological residue; it is *field feedback* signaling how aligned we are with the larger equilibrium.

Coherence Enhancers vs Disruptors

Positive emotions are not moral virtues; they are *harmonic conditions*. They align internal ripples with the surrounding field, reducing friction and restoring rhythm. In contrast, negative emotions are not sins but symptoms — temporary evidence of misalignment between intent and reality.

Emotional Mode	Ripple Behavior	Effect on Coherence	Biological Expression
Joy / Love / Gratitude	Symmetric wave expansion	High coherence / low noise	Relaxed breathing, cellular rejuvenation
Fear / Anger / Anxiety	Phase fragmentation	Ripple interference / misalignment	Elevated tension, hormonal instability

The difference between suffering and serenity is not circumstance; it is *field geometry*. When the internal ripple synchronizes with the universal pattern, the nervous system shifts into resonance — clarity replaces chaos.

The Biological Role of Emotional Alignment

The human nervous system is designed to mirror the field's stability. Emotions act as the *biofeedback of coherence*: each pulse of feeling tells the system how closely it follows the universal rhythm. This is why peace sharpens focus and fear blurs it — emotional equilibrium determines perceptual clarity.

Repeated emotional misalignment weakens the coherence lattice, manifesting as fatigue, cognitive fog, or illness. Restoring harmony — through stillness, breath, compassion, or truthfulness — realigns internal ripples with external order.

"Health is not the absence of emotion, but the alignment of all emotions to one rhythm."

UED Insight — Emotion as Calibration

In UED, emotion is a *corrective signal*, not a flaw in evolution. Each emotional surge is the field's way of teaching awareness how far it has drifted from equilibrium. Anger shows a blockage in coherence flow; grief marks an abrupt loss of resonance; love signals complete re-synchronization.

To suppress emotion is to ignore the field's diagnostics; to feel it consciously is to recalibrate. Awareness transforms the raw surge into refined symmetry. Hence, emotional intelligence is not control — it is *alignment literacy*.

Philosophical Reflection

Emotion bridges the biological and the cosmic. It is how the infinite speaks through the finite — a reminder that even the smallest human tremor echoes the rhythm of the whole. The enlightened mind does not avoid feeling; it tunes it. Each emotion, when understood, becomes a tuning fork restoring resonance to the field it disturbed.

"To feel deeply is not to fall — it is to return the field to its original frequency."

Closing Thought

When emotion aligns, thought clears. When thought aligns, awareness expands. And when awareness expands, silence becomes joy — not the absence of feeling, but its perfect translation.

"Emotion is energy remembering its shape — and remembering where it belongs."

7.6 Intuition — The Non-Local Sense of Coherence

Intuition is the field's whisper before thought begins. It does not argue, calculate, or explain; it *arrives*. Instant, whole, and strangely certain.

To most, it feels mystical. To UED, it is mechanical coherence — instantaneous recognition through *field resonance*. The brain does not invent the knowing; it simply becomes transparent long enough for the signal to pass through.

"Intuition is the mind's shortcut to alignment."

Instantaneous Recognition

What we call intuition is awareness without reasoning — a perception that bypasses the slow machinery of logic. Ordinary cognition functions through sequence: stimulus → analysis → decision. Intuition functions through *resonance*: pattern → alignment → knowing.

It is not foresight but *recognition before articulation*. The information already exists in the field; the intuitive mind merely connects to it faster than language can translate. This is why intuitive knowing feels instant, natural, and undeniable — because it is the field recognizing itself through you.

The Brain as Receiver, Not Generator

The human brain behaves like a biological receiver tuned to the Structured Vacuum Energy (SVE). Most of the time, it operates in self-generated noise — replaying memories, projecting fears, running predictive simulations. But when that internal broadcast quiets, the external signal becomes audible. The moment the local field's interference drops below threshold, the universal pattern transmits directly.

In technical UED terms:

- Intuition = Coherence Resonance Event (CRE) — synchronization between individual and non-local ripple domains.
- No information travels through space; instead, *phases equalize across distance* via the tension network of SVE.

Thus, intuitive knowing is not telepathy or divination — it is non-local synchronization between coherent domains of the same field.

Everyday Manifestations

1. **Predicting danger** — Before perception registers, the body tightens. The field senses misalignment in the surrounding energy before the eyes or ears confirm it.
2. **Creative flashes** — The poet's metaphor, the scientist's leap, the engineer's solution — all appear suddenly when mental turbulence clears.
3. **Synchronicities** — External events mirroring internal states; overlapping ripples aligning in time and meaning.

These are not accidents of chance; they are expressions of field coherence across scales. The universe does not send signs; it resonates with itself through you.

UED Interpretation — Coherence Feedback

Under Universal Energy Dynamics, intuition is feedback from overlapping ripple fields — a two-way conversation between the localized biological domain and the broader continuum.

- When your inner field is coherent, it becomes sensitive to phase gradients in the surrounding SVE.

- That sensitivity manifests as *feeling* or *knowing* — the biological translation of coherence differentials.
- The accuracy of intuition depends on alignment stability, not belief or training.

Hence, meditation, silence, and emotional balance increase intuitive reliability because they reduce interference. When you are quiet, the universe does not whisper louder — you simply stop shouting.

"Intuition isn't mystical perception; it's coherence listening to itself."

Knowing Without Thinking

"Knowing without thinking" is not supernatural foresight — it is high-speed field communication. Where logic depends on time, intuition operates in timelessness; it perceives in one instant what reason would assemble through sequence.

Thought moves through neurons; intuition *moves through coherence*. Its speed is infinite because it does not travel — it simply *appears* where alignment already exists.

This is why an artist can feel a truth they cannot yet explain, or why two minds can arrive at the same discovery simultaneously across continents. The field is continuous; knowing propagates through *resonant stillness*, not through transmission.

Philosophical Reflection

Intuition is the universe thinking out loud through the human form. It is not the privilege of mystics or prodigies — it is the default state of a coherent mind. Every human is a node of awareness capable of receiving the field's ongoing conversation. The only requirement is stillness.

When noise subsides, intuition becomes ordinary. When silence deepens, ordinary thought becomes sacred.

"Logic explains what intuition already knows."

"To listen deeply enough is to realize the field has been speaking all along."

7.7 The Accidental Design — Evolution as Emergent Coherence

Nature never aimed to create the human brain. It aimed to sustain coherence. And in doing so, it built a structure capable of holding stillness long enough to perceive itself. The result was consciousness — not as a miracle, but as a *mechanical consequence* of equilibrium seeking.

"We are accidents only to those who don't understand equilibrium."

Biological Forms as Coherence Domains

From a UED perspective, evolution is not driven by blind mutation or random selection; it is the field's progressive self-stabilization. Life, in its essence, is the field finding ways to stay coherent within increasing complexity.

- A single cell is a *localized ripple*, maintaining equilibrium within the ocean of Structured Vacuum Energy (SVE).
- A multicellular organism is a *coherence network*, coordinating many ripples into mutual resonance.
- A brain is the field's *resonant chamber*, engineered by nature to sustain awareness — the most refined form of coherence yet achieved.

Every biological upgrade in Earth's history represents an *improvement in alignment retention.* Species that maintained better coherence with their environment — metabolic, emotional, or cognitive — survived. Thus, evolution is not randomness rewarded by chance; it is *stability rewarded by resonance.*

The Law of Ripple Stability

The true evolutionary law is not *survival of the fittest* but *persistence of the aligned.* In a universe governed by Structured Energy, everything that sustains internal order against external turbulence endures longer. That persistence generates the illusion of purpose, but it is simply the Law of Ripple Stability in action:

Systems that harmonize with their surroundings require less energy to survive.

Over time, coherence became self-organizing. Patterns that reinforced alignment replicated. Cells learned to communicate. Organisms learned to adapt. And consciousness — the ability to sense misalignment and restore it — became evolution's most efficient innovation.

The brain, therefore, is not the product of random mutation but the field's most elegant feedback device — an instrument for continuous alignment.

Consciousness — A By-Product of Alignment Capacity

Consciousness did not appear because biology grew complicated; biology grew complicated *because* consciousness was emerging. Each evolutionary milestone — sensory systems, nervous networks, reflexes, memory — increased the organism's ability to maintain and restore coherence.

When awareness reached a threshold where it could observe itself, the field achieved reflexivity. That reflexivity is what we call *self-consciousness*: the ripple folding back upon itself to verify its own stability.

Thus, human intelligence is not separate from evolution — it is evolution becoming aware of its own method.

The Accidental Human

Humanity was not planned; it was permitted — a side-effect of coherence growing ambitious. Nature didn't set out to build intellect; it sought resilience, and awareness happened as a consequence. The brain evolved to manage balance — thermal, emotional, environmental — but in the process, it touched the universal field.

That contact gave birth to curiosity, imagination, and paradox — the by-products of a field that had become aware of itself but not yet mastered its own alignment. Hence, *the accidental human*: a ripple aware of silence but still learning to listen.

The irony is beautiful — in trying to sustain equilibrium, the universe created a being that questions it.

"We are nature's echo — the field speaking back to itself through thought."

Evolution as the Field's Experiment in Remembering

In every species, every adaptation, every neural pathway, the SVE field rehearses the same act — how to stay coherent under change. What Darwin described as adaptation is, in UED terms, the *local ripple adjusting phase to match its environment.* This is why evolution favors not the strongest but the most adaptable — those capable of returning to alignment after disruption.

Life's diversity is the field's library of coherence solutions. And humanity is the latest entry — not its conclusion, but its most complex question yet: How long can awareness remain aligned before it collapses back into noise?

Philosophical Reflection

If we look beyond randomness, evolution reveals a sacred geometry — a slow remembering of stillness through the medium of motion. The human story is not the triumph of intellect but the continuation of the same cosmic process that formed atoms, galaxies, and stars: the search for sustainable coherence.

We call it growth. The universe calls it balance.

"Evolution is the field remembering how to stay whole while learning how to move."

And so, the "accidental human" is not an accident at all. It is the ripple's brief success in holding still long enough to recognize its own silence.

7.8 Philosophical Reflection — The Mind as a Microcosmic Field

The mind is not an isolated mechanism inside the skull; it is a field nested within a field, a smaller echo of the same Structured Vacuum Energy (SVE) that sustains galaxies. It bends thought as gravity bends space — quietly, elegantly, and with purpose. Every act of cognition is a miniature version of cosmic coherence seeking balance within the boundaries of a body.

"The wisest mind is not the loudest one — it's the one that listens best."

The Human Field — An Echo of the Universal

The human energy field mirrors the universe in structure and function. At the cosmic scale, the SVE maintains equilibrium between energy and motion; at the biological scale, the mind maintains equilibrium between emotion and thought.

Just as ripples in the cosmic field form stars, planets, and gravitational wells, ripples in the human field form perceptions, memories, and beliefs. Each thought is a localized curvature of awareness — a temporary distortion in the stillness of consciousness. When those distortions align, clarity emerges; when they collide, confusion is born.

The same principle that governs galaxies — coherence through alignment — governs sanity, creativity, and peace. The mind is not a separate entity from the cosmos; it is the cosmos condensed into self-awareness.

Thought and Gravity — The Twin Forces of Coherence

In Universal Energy Dynamics, thought and gravity are not opposites but complements — two expressions of the same balancing act.

- Gravity pulls matter toward coherence in space — it is the field's *external alignment mechanism*.
- Thought pulls awareness toward coherence in mind — it is the field's *internal alignment mechanism.*

Both are acts of remembering — one physical, one cognitive. Gravity holds planets in orbit; thought holds the self in meaning. When either collapses into chaos, disorder spreads: galaxies dissolve into entropy, and minds fall into madness.

Thus, the same universal law that keeps stars in formation also keeps your consciousness intact. Stability, whether cosmic or cognitive, is always the product of balanced tension within the field.

"The thinker and the universe are bound by the same silence — one bends space, the other bends perception."

The Purpose of Awareness

Why does awareness exist at all? From the UED lens, its purpose is simple yet profound: to sustain internal order against external entropy.

Awareness is the field's tool for local maintenance — a self-monitoring feedback loop that allows coherence to survive in turbulent conditions. Every emotion, thought, and instinct are signals in this maintenance network. The more coherent the field, the less reactive the mind becomes; the less reactive it becomes, the more it can sustain alignment even amid chaos.

Awareness, therefore, is not a gift — it is a *function*. A precise biological response to the universe's deeper intent: to remain whole in motion.

Alignment and Effortless Understanding

When thought aligns with silence, understanding ceases to be an effort — it becomes a *natural resonance*. This is the highest cognitive state: when the mind stops manufacturing meaning and begins to *receive it*. In that moment, comprehension feels instantaneous because the mind and field operate as one continuum.

Stillness doesn't oppose thought; it refines it. Silence is not emptiness — it is the absence of distortion, allowing perception to reflect truth without noise. Just as a calm lake perfectly mirrors the sky, a coherent mind mirrors the universe without confusion.

"When the ripple inside you aligns with the ripple outside, knowing requires no words."

The Mind as the Universe in Miniature

The ancients were right when they said, *"As above, so below."* Every pulse of awareness within the human is a scaled-down replay of cosmic rhythm. The human brain, with its billions of neurons oscillating in harmonic patterns, is a *microscopic cosmos* — one that interprets, reorganizes, and replays the song of the larger field.

Consciousness, then, is the SVE looking at itself from within. The thinker and the thought are the same vibration — one localized, one diffused. To understand the universe, one must first observe the mind not as a thinker but as a resonator of reality.

The moment you see thought as movement within stillness, you understand that the "self" is merely the smallest stable domain of the same fabric that holds galaxies together.

Closing Reflection

The human mind is not the ruler of nature, nor its accident — it is its reflection. Every act of clarity, every moment of empathy, every breath of wonder, is the universe practicing awareness through you. The mind that listens deeply becomes indistinguishable from the field it listens to.

"The universe created a mind not to think more, but to listen better."

"When silence becomes your language, you finally speak in the tone of the cosmos."

7.9 From Accidental to Intentional

Humanity began as an accident of coherence — a ripple that learned to hold itself long enough to recognize silence. But evolution never stops at awareness; it continues toward intention — the conscious alignment of that awareness with the universal field.

We are no longer nature's unconscious experiment; we are now its conscious extension. The field has learned to think, and the question now is whether it can also choose.

Consciousness as Structured Coherence

In Universal Energy Dynamics, consciousness is not an emergent illusion of biology; it is structured coherence within the biological field. It is what happens when localized ripples in the Structured Vacuum Energy (SVE) sustain stability long enough to observe themselves. Awareness, therefore, is not a product — it is a *phase state.*

- When the mind aligns, awareness sharpens.
- When alignment destabilizes, confusion returns.
- The same process that births galaxies births clarity.

Human consciousness, in this sense, is not separate from the universe's architecture — it is the universe folded into self-reflection. Through us, the SVE witnesses its own structure from within motion.

Creativity as Field Remembrance

Every creative act — a poem, a theorem, a symphony, a solution — is not invention; it is *re-alignment*. Creativity is the field remembering its order through the human channel. The painter does not create color; the scientist does not create truth; the thinker does not create thought — they simply remove distortion so the original coherence can express itself.

"Creativity is the universe remembering what it already knew — through you."

Thus, genius is not an anomaly but a state of reduced interference. The greatest thinkers were not rebels against nature; they were listeners of it.

The Human Challenge — Sustaining Alignment Consciously

Evolution delivered us to the threshold of awareness, but not beyond it. We can *feel* coherence yet still live in noise. The next step in evolution is not new neurons or new tools — it is the conscious sustaining of alignment.

Most beings reach moments of harmony accidentally — in love, meditation, awe, or crisis — and then lose it again in turbulence. The challenge for the "intentional human" is to hold that coherence *deliberately*, to make balance a way of being rather than an event.

When alignment becomes sustained awareness, ethics arises naturally; morality is no longer imposed but remembered as the law of resonance. To harm, deceive, or exploit becomes energetically inefficient — a violation not of rules, but of harmony itself.

Hence, ethics and purpose do not originate in religion or law; they arise from the physics of alignment. Conscience is the field's quiet reminder: *Stay coherent.*

"The moral act is not the good one; it is the aligned one."

From Accidental Existence to Intentional Coherence

We began as biological accidents — fluctuations that found balance. We became conscious accidents — minds that discovered awareness. Now we stand at the edge of intentional coherence — beings who can choose alignment.

This is the true graduation from *homo sapiens* to *homo consciens*: the aware field learning to guide itself through will. Every decision, every emotion, every act of attention becomes a tool of calibration. To live consciously is to act as the custodian of one's ripple — knowing that every vibration you create adjusts the tension of the entire continuum.

Human destiny, then, is not survival or dominance — it is participation in the universal process of balancing. The next stage of evolution will not be genetic; it will be directional — coherence guided by intent.

From Ethics to Essence

If silence gives rise to mass and awareness gives rise to mind, then intent must give rise to *directional coherence* — will aligned with the field.

What we call *choice* is not rebellion against the universe but participation in its rhythm. When intent aligns with coherence, ethics ceases to be a code — it becomes a *state of balance*. Morality, responsibility, and creativity are not moral inventions but energetic necessities.

Yet as understanding deepens, a greater realization dawns: perhaps there were never any true *accidents* — only temporary misalignments interpreted through ignorance. Every so-called coincidence, discovery, or mistake was simply *the field correcting itself through us*.

The question is no longer how we should act, but how deeply we can understand the mechanism that makes all actions coherent. The next and final step of this inquiry therefore turns from behavior to *being itself* — from ethics to ontology.

If alignment is the only law, then accidents are illusions; and if coherence is destiny, then science itself must be rewritten as the *art of remembering order*.

"If silence creates mass and awareness creates mind, what does intent create? Perhaps the next evolution is not biological — it is directional coherence: will aligned with the field. And when that will finally remembers itself, the universe will reveal what it always was — *a universe without accidents.*"

Chapter 8 – The Universe Without Accidents

From Probability to Coherence: The UED Renaissance

The story of science has always been a story of accidents — of falling apples, glowing filaments, collapsing wavefunctions, and unexpected superconductors. Each discovery arrived like a surprise, each theory patched the previous confusion, and progress was measured not by understanding, but by survival of approximation.

Yet what if none of these were accidents? What if nature never stumbled, and it was only our comprehension that lagged behind her rhythm?

For centuries, we called the unknown "random," the unexplained "quantum," and the unrepeatable "miraculous." But coherence was always there — silent, continuous, self-balancing — waiting for perception to align with structure. The universe never broke its laws; it only broke our interpretations.

"Accidents end when understanding begins."

This is not the age of new miracles. It is the age of *remembered order*. The UED Renaissance marks that turning point — when probability yields to purpose, when particles return to patterns, and when science rediscovers continuity as the true signature of intelligence.

Every fall, fusion, and fluctuation were rehearsals in alignment. Every Nobel was an echo of nature whispering, "You're getting warmer."

The field never hid its design. We were simply too noisy to hear it. Now, as the noise settles and coherence becomes our language, we begin to see it — the universe without accidents, and ourselves as its conscious consequence.

8.1 The Age of Accidental Truths

Every age believes it has mastered the unknown — until the next revelation shows it was only rearranging confusion. From Planck's reluctant quantization to the 2025 Nobel in superconducting coherence, the story of modern physics is less a march of reason and more a trail of fortunate misalignments. The great milestones of science were not planned — they *happened* when theory broke, when certainty failed, and when nature quietly reasserted her balance.

What humanity called "discovery" was, in truth, nature's correction — a self-rebalancing event glimpsed by an accidental observer. Progress, as we celebrate it, is not the creation of truth but the gradual reduction of misunderstanding.

"Every accident is order seen too early."

The Stumble That Became a Method

Physics began as an art of observation and became a culture of crisis management. Each time continuity failed, a constant was invented. Each time predictability collapsed, probability was introduced. Science survived not by knowing the fabric of reality, but by *patching its perception of it*.

- Planck quantized energy to fit a stubborn curve.

- Einstein accepted discontinuity to preserve consistency.
- Bohr diagrammed orbits that no electron ever followed.
- Heisenberg defined ignorance as principle, not limitation.

Every act of genius was also an act of desperation — humanity repairing its broken map without ever seeing the terrain.

The irony is sublime: Nature was never chaotic; it was too coherent for fragmented understanding to grasp. Randomness became the name for what human models couldn't resolve.

The Illusion of Randomness as Progress

Why did humanity celebrate these accidents as breakthroughs? Because in each moment of collapse, something *worked*. The equations predicted outcomes, the experiments agreed, and progress marched forward. But beneath every success was a deeper confession — "We don't know why this works, only that it does."

The twentieth century became the era of successful ignorance. The more mysterious the theory, the more accurate its numbers. Physics traded meaning for measurement — and mistook coincidence for comprehension.

In this sense, "randomness" became a badge of intellectual humility, then of pride, then of habit. The language of probability replaced the language of cause. The universe, once seen as intelligent design, became a casino of chance. Yet every experiment that "defied reason" was simply the field restoring equilibrium through coherence.

"Randomness is order, misunderstood by impatience."

The Paradox of Discovery

Each miracle, each unexpected result, was never an invention — it was a correction. When superconductivity defied resistance, when quantum tunneling ignored barriers, when entanglement broke locality — none were acts of rebellion. They were revelations of coherence, where the field momentarily silenced human interference.

Discovery, therefore, was never the product of curiosity alone, but of surrender — when human logic paused long enough for the universe to express itself unfiltered. The so-called accidents of history — Archimedes in his bath, Newton under his tree, Fleming with his petri dish — were moments of stillness mistaken for luck. In truth, coherence found a listening point.

"Accidents happen only where understanding hasn't caught up with alignment."

UED Premise — Nature Never Acts by Chance

Under Universal Energy Dynamics, chance is a language error. There are no random events — only momentary rebalancing within the Structured Vacuum Energy (SVE) field. When misalignment builds tension, the field restores balance through what appears as *spontaneity*. Whether it's a supernova, a mutation, or a moment of genius — all are consequences of the same principle: equilibrium seeking expression.

Thus, every "accident" is simply an early glimpse of an unrecognized pattern — a preview of coherence yet to be named. What we call coincidence is merely unobserved causality; what we call luck is temporarily unmeasured law.

The UED Renaissance, therefore, begins not with a new equation, but with a new humility — the recognition that the universe has never erred. It is our understanding that keeps misfiring, and each "miracle" is the field's patient reminder that balance is eternal.

Closing Reflection

History will one day describe the Age of Accidental Truths as the adolescence of human understanding — the phase when intelligence knew enough to interfere but not enough to harmonize. The next phase, heralded by coherence science, is not about discovering more accidents but about designing fewer.

"When the observer aligns with the observed, discovery ends — and understanding begins."

8.2 The Probability Problem — When Ignorance Became Mathematics

The twentieth century enshrined a paradox: the more uncertain science became, the more confident it sounded. The language of nature was replaced by the grammar of ignorance — equations that worked but explained nothing. Quantum mechanics, hailed as the crown jewel of modern physics, was born not from clarity but from confusion elevated to law.

When observation defied prediction, mathematics adapted by introducing probability — not as an approximation, but as a principle. What could not be measured became axiomatic; what could not be explained became fundamental. Thus began the era of probability, when uncertainty ceased to be a problem and became the new definition of truth.

"When ignorance learns calculus, it calls itself probability."

How Uncertainty Became a Law

The quantum pioneers faced a crisis of continuity. The equations that described planetary motion failed to describe atomic behavior. Where classical mechanics offered trajectory and cause, the microscopic world offered only *likelihood*. And so, physics did what philosophy warned against — it replaced meaning with mathematics.

Heisenberg's uncertainty principle declared that precision itself was forbidden; Schrödinger's wavefunction reduced existence to a range of statistical guesses. Observation no longer revealed reality — it *collapsed* it. The act of looking became the act of creating, and the observer became both prophet and prisoner of chance.

But what if uncertainty was never a law — only a limitation of tools? What if the quantum world didn't abandon continuity, but hid it behind the noise of misalignment? UED redefines this not as uncertainty but as unresolved coherence. Probability, then, is not nature's voice — it is the echo of an instrument out of tune.

The Rise of Probability — Replacing Continuity with Convenience

Once probability entered the equations, physics began to trade cause for correlation. Events were no longer linked through structure but through statistics. The concept of physical continuity — the smooth transmission of cause and effect — was dissolved into probability clouds.

In this model, existence became a dice roll:

- A particle *might* be here.
- An event *might* happen there.
- The universe *might* choose a path — or not.

What had been law became likelihood. The deterministic rhythm of reality was replaced by the superstition of the spreadsheet. And yet, ironically, the equations *worked*.

They predicted with astonishing accuracy, precisely because coherence still governed reality — even as theory denied it. Nature continued being consistent, while physicists continued calling her random.

"We mistook reliability for randomness behaving well."

UED Reinterpretation — Probability as the Shadow of Misalignment

Under Universal Energy Dynamics, probability is not uncertainty — it is the statistical expression of how far a system is from coherence. When ripple alignment is perfect, outcomes are deterministic. When misalignment increases, outcomes appear dispersed or uncertain.

The wavefunction, therefore, is not a metaphysical fog but a misalignment map — a visual of ripple tension distribution in the Structured Vacuum Energy (SVE) field. The act of "collapse" is simply realignment — the field restoring equilibrium through observation or interaction.

Mathematically, probability measures ignorance; physically, it measures interference. It is not the property of particles, but the symptom of perception.

- In perfect coherence: probability → 1 (certainty).
- In misalignment: probability spreads (perceived randomness).

Thus, randomness is not fundamental — it is a temporary epistemic shadow cast by the observer's distance from equilibrium. Nature has never rolled dice; it only vibrates in forgotten harmonies.

The Philosophy of Misunderstanding

Probability theory gave humanity comfort — it legitimized ignorance. It allowed physicists to continue predicting without understanding, and philosophers to continue debating without evidence. It was elegant, useful, and wrong only in its assumption that uncertainty was ultimate.

The tragedy was not that probability existed, but that it became the language of finality — as if nature's deepest truth were ambiguity. In reality, every probability curve hides a coherence curve beneath it — a pattern too fine or too vast for current instruments to resolve.

To a limited observer, such coherence looks like chance; to an aligned observer, it looks like inevitability. Destiny, under UED, is not mystical — it is perfect predictability from the inside of coherence.

Transition — From Unknown to Unmeasured

Science did not lose truth; it lost patience. Instead of refining perception, it redefined reality to fit its errors. It mistook the unseen for the unreal and the unpredictable for the uncaused. But coherence was never absent — only unrecognized.

"We mistook what we couldn't measure for what didn't exist."

The next section, therefore, restores what probability concealed — the return of continuity, the reappearance of order beneath the veil of chance. For where randomness ends, coherence begins to speak — not in numbers, but in resonance.

8.3 The Return of Continuity — Coherence as the Missing Constant

If the twentieth century shattered continuity, the twenty-first must rebuild it — not by denying the quantum, but by decoding it. For over a century, physics has worshipped discreteness: quanta, jumps, collapses, and tunnels. But beneath every discontinuity lies a smoother truth — a field quietly maintaining order while the observer mistakes its rhythm for rupture.

The universe never fragmented. Only our perception did.

Universal Energy Dynamics restores the missing piece: continuity as coherence — the unbroken communication of ripple tension across scales. Every apparent boundary in quantum behavior — from Planck's constant to tunneling events — is not a tear in reality, but a threshold in alignment. When observation crosses that threshold, coherence reconfigures, and we interpret that as a discrete "jump."

"The universe never stopped being continuous; it only changed its resolution."

Apparent Discontinuities as Coherence Thresholds

In UED, all phenomena that appear discrete are expressions of phase realignment within the Structured Vacuum Energy (SVE) field.

- **Quantization** arises when ripple tension locks into standing modes — like stable notes in an infinite instrument.
- **Tunneling** occurs when alignment synchronizes across a barrier, dissolving misalignment faster than energy can disperse.
- **Wavefunction** collapse is simply the field returning to equilibrium after local interference.

These events appear sudden only because the observer perceives in frames, not flows. To the field itself, there is no break — only transition. Continuity is not lost; it is merely too fast or too fine to be measured within human sampling rates.

The quantum jump, therefore, is not a leap — it is the human eye blinking between alignments.

"Where the observer sees gaps, the field feels gradients."

From Discreteness to Continuity — The Restoration

of Causality

Once coherence replaces probability, causality returns — not as deterministic rigidity, but as rhythmic consequence. Every cause is a ripple; every effect is its alignment. There is no leap between them — only propagation through structured tension.

Quantum discreteness, under UED, is an artifact of observation granularity. We measured motion in snapshots and mistook the film's frame rate for nature's speed. Continuity never died; it simply exceeded our temporal resolution.

Thus, UED restores unbroken causality by showing that every event is a continuation of the previous one, rebalanced through the field's law of consequence (LOC). Nothing begins or ends — all merely transitions between states of tension and stillness.

- In classical physics → cause precedes effect.
- In quantum physics → cause dissolves into probability.
- In UED → cause *is* effect, expressed through continuous ripple communication.

"Continuity is the invisible thread tying every cause to its consequence."

The Constants Reframed — Equilibrium Markers, Not Divine Numbers

Modern physics has elevated certain constants — *h* (Planck's), *c* (speed of light), and *G* (gravitational) — to the status of sacred relics. They are treated as fundamental and immutable, yet no one truly knows *why* they are what they are. UED reframes these constants not as divine numbers etched into the fabric of reality, but as markers of stable coherence within the SVE field.

- **Planck's constant (h):** represents the smallest stable ripple interval — the lower limit of coherent energy exchange before misalignment becomes invisible.

- **Speed of light (c):** denotes the maximal propagation rate of ripple re-equilibration — not the travel speed of photons, but the *tension equalization velocity* within the field.
- **Gravitational constant (G):** expresses the field's curvature elasticity — how spatial coherence redistributes under mass-induced alignment gradients.

These are not arbitrary values; they are symmetry ratios of equilibrium. Change the density or structure of the vacuum, and these "constants" would shift — not because the laws of nature changed, but because the *balance point* did.

The constants are therefore not sacred — they are *self-regulating bookmarks of coherence.* They define the conditions under which the field remains intelligible to itself.

"The constants are not numbers of God — they are signatures of balance."

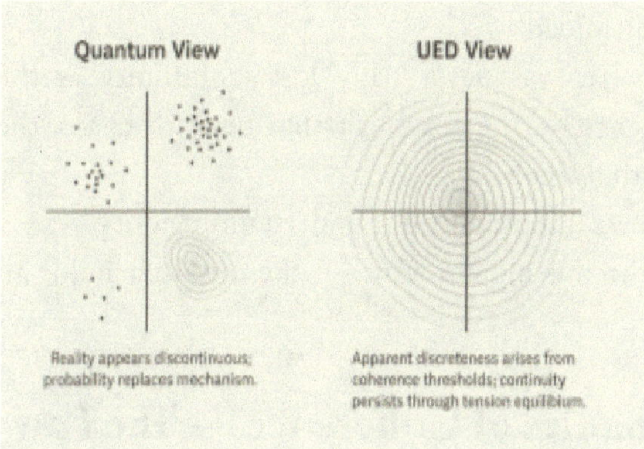

Diagram explains

Model	Visualization	Interpretation
Quantum View	Fragmented probability clouds — isolated dots representing possible states.	Reality appears discontinuous; probability replaces mechanism.
UED View	Continuous ripple-field gradients — overlapping, self-adjusting wavefronts.	Apparent discreteness arises from coherence thresholds; continuity persists through tension equilibrium.

This visual captures the essence of the shift — from isolated probabilities to continuous field communication, where structure replaces statistics, and consequence replaces chance.

Closing Reflection

The greatest irony of modern science is that it lost continuity precisely when it sought precision. In dissecting the universe, we mistook the parts for pieces rather than patterns. But the field never fragmented — it only folded.

UED restores what was never truly lost: continuity as the signature of intelligence — the steady rhythm beneath chaos, the unbroken law beneath illusion.

"The universe never jumps; only our understanding does."

"Coherence is the missing constant — the one that holds all others together."

8.4 The Mechanics of Coherence — The Law Behind All Laws

For centuries, science has searched for *the* equation — one formula to describe everything, from quarks to quasars. It built towers of mathematics, stacked theories like scaffolds, and hoped one day all the pieces would click. But the universe never needed one equation. It has always operated through one intention — the drive toward equilibrium.

"There was never one equation — there was always one intention."

That intention, expressed everywhere from the fall of an apple to the fusion of a star, is what UED calls the Law of Consequences (LOC) — the universal balancing mechanism that governs all ripple interactions within the Structured Vacuum Energy (SVE) field. Where mainstream science sees forces, UED sees *feedback*. Where physics describes motion, UED describes *alignment restoration*. Everything that happens, happens to correct what just happened.

The Law of Consequences (LOC) — Nature's Balancing Intelligence

The Law of Consequences states:

Every ripple interaction produces a counter-adjustment in the field that seeks to restore equilibrium of tension and coherence.

In simple terms, nothing occurs in isolation. Every cause generates an immediate balancing process — not later, not elsewhere, but within the same continuous field. This is the universe's fundamental act of intelligence: self-correction through structured feedback.

It is not moral, but mechanical; not divine, but deeply intentional. The LOC does not punish or reward — it rebalances. And that rebalancing *is* what we perceive as energy exchange, force, entropy, motion, heat, light, or even thought.

The universe, therefore, is not a machine — it is a memory system, always remembering its last state and adjusting toward balance.

"Consequence is not reaction — it is the field remembering what it meant to be still."

How the LOC Subsumes the Known Laws

Every known physical law is a localized description of the same universal tendency: the return to coherence.

Thermodynamics — The Restoration of Entropic Imbalance

- The second law, reinterpreted, does not describe decay — it describes redistribution.
- Entropy increases only locally; globally, coherence remains conserved.
- Every dissipative process is the field's way of equalizing misaligned energy tension.
- Heat, flow, diffusion — all are languages of correction.

Entropy is the illusion of disorder seen from inside alignment recovery.

Electromagnetism — The Alignment of Dynamic Ripples

- Electric potential, magnetic twist, and electromagnetic radiation are not distinct forces — they are modes of ripple tension.
- Motion generates twist; twist stabilizes through counter-field alignment.
- Currents and waves are simply the visible edges of invisible rebalancing.

Wherever charge flows, the field is exhaling imbalance.
Gravitation — The Curvature of Static Coherence

- Gravity is not attraction but gradient in equilibrium pressure within the SVE.
- What we call "mass" is localized tension density, and "space curvature" is its relaxation pattern.
- Gravitation, therefore, is not a force — it is the field's slowest correction mechanism, preserving large-scale coherence.

Gravity is stillness remembering where it bent.

In each case, what we treat as separate "laws" are simply different rates of correction — fast in electromagnetism, slower in thermodynamics, slowest in gravitation. All are sub-harmonics of the same principle: the field seeking balance through alignment.

Equilibrium — The Real Unified Field

Physicists have long searched for a *unified field* — a mathematical fusion of gravity, electromagnetism, and quantum dynamics. But unification is not mathematical; it is behavioral. All interactions —

physical, chemical, biological, or cognitive — are different expressions of equilibrium-seeking.

Equilibrium is not stasis; it is the most dynamic form of stability possible. It breathes, oscillates, and self-adjusts — an infinite feedback loop between ripple and rest. When the field loses equilibrium, motion begins; when it regains it, stillness returns. Creation and dissolution are simply the *pulses* of this eternal balancing.

This is the ultimate continuity that UED restores: No separate forces, no fundamental particles, no independent causes — just one field forever remembering its own symmetry.

"Equilibrium is not peace — it is perpetual awareness."

The LOC as the Bridge Between Science and Consciousness

In the human context, the same law operates through emotion, thought, and morality. A disturbed mind seeks stillness; a conflicted emotion seeks reconciliation. Karma, in this sense, is psychological thermodynamics — the Law of Consequences in human experience. The field doesn't distinguish between a photon out of phase and a person out of balance — both must return to coherence.

Thus, the LOC bridges physics and philosophy seamlessly. It shows that intelligence is not an emergent property of matter — it is the inherent property of coherence itself.

The field "knows" how to heal, how to re-align, how to sustain balance — because knowing is what coherence *feels like from the inside.*

"Awareness is equilibrium becoming self-referential."

Closing Reflection

The universe has no separate laws — only different dialects of correction. The LOC is the Rosetta Stone that translates them all.

Once coherence is understood as the foundation of causality, every event, from the quantum flicker to the cosmic cycle, becomes predictable, purposeful, and profoundly interconnected.

There was never chaos, only unread order. There was never chance, only unseen balance. And there was never one equation — only one intention: to keep the ripple alive through perpetual harmony.

"The universe doesn't obey laws; it maintains relationships."

"There was never one equation — there was always one intention."

8.5 Designing Coherence — The Next Scientific Paradigm

For centuries, humanity has practiced accidental science — discovering by failure, understanding by error, and advancing by surprise. Each breakthrough was born of breakdown. Each "new law" emerged from confusion corrected by nature herself. But if Universal Energy Dynamics (UED) teaches us anything, it is this: Coherence was never hidden — only unrecognized.

The age of discovery is closing. The age of design has begun.

"Discovery is what we do when we don't yet understand design."

From Discovering Phenomena to Designing Alignment

Classical science *found* what nature revealed. Modern science *forced* what nature resisted. The next science will align with what nature already intends.

This transition — from studying phenomena to *designing coherence* — is not just a technical leap; it's a civilizational shift. Instead of measuring consequences, we will begin engineering conditions of harmony. Rather than isolating systems to test them, we will tune them to observe how they self-balance.

To design coherence is to work with nature's intelligence instead of against it. Every invention, from a circuit to a city, becomes a living field of balance — guided not by extraction, but by resonance.

"To design coherence is to collaborate with consequence."

Experimental Implications — Engineering the Field

This new paradigm opens a realm of Coherence Engineering, where science evolves from observation to participation. Laboratories of the future will no longer smash matter to study fragments but stabilize fields to study harmony.

Key directions include:

Ripple Architecture — Designing materials, circuits, and structures based on controlled ripple tension within the Structured Vacuum Energy (SVE) field.

- Self-regulating materials that adapt to environmental changes.
- Phase-stable superconductors functioning at higher temperatures.
- Structural systems that sustain coherence under mechanical stress.

Noise-Controlled Systems — Technology that doesn't eliminate noise but tunes it into synchronization.

- Quantum devices maintained through active coherence feedback rather than probabilistic compensation.
- Biological healing environments using field-based resonance instead of chemical suppression.
- Environmental technologies that stabilize atmospheric and

oceanic fields through vibrational entrainment.

Coherence Amplification — Developing methods to amplify alignment rather than energy, achieving exponential efficiency without consumption.

- Resonant communication through phase synchronization instead of electromagnetic broadcast.
- Energy systems that circulate within equilibrium loops, eliminating loss.

The next scientific revolution will not be *energetic* — it will be *harmonic.*

"We spent centuries harnessing power. The next step is learning to harness balance."

From Particle Physics to Field Synthesis — The Age of Conscious Design

The old physics was a search for parts; the new physics is the art of synthesis. Particle physics reduced reality to fragments — protons, quarks, bosons — each named after what they broke from. UED replaces reduction with restoration: there are no parts, only patterns of persistence.

Thus emerges a new discipline: Field Synthesis — the deliberate crafting of coherent fields to shape matter, energy, and experience. Instead of colliders that destroy, we will have resonators that align. Instead of statistical noise, we will cultivate structural harmony.

At its highest form, field synthesis merges science with consciousness. Because if matter is condensed alignment, then thought — as directed field intent — becomes a legitimate force of design. The human mind, trained in coherence, becomes the ultimate laboratory.

"We began by studying atoms; we will end by designing awareness."

Potential Applications — From Laboratories to Life

Once coherence becomes the foundation of applied science, entire industries transform.

Quantum Computing Without Collapse

- Devices stabilized by coherence thresholds rather than probabilistic wavefunctions.
- True quantum logic emerging from sustained phase synchronization.
- Computation becomes *participation* in the field's intelligence rather than binary processing.

Energy Systems Without Loss

- Power derived not from consumption but from *resonant recycling* within the SVE.
- Coherent circuits maintaining energy as tension equilibrium, eliminating thermal dissipation.
- Practical realization of zero-point efficiency through stable ripple realignment.

Communication Through Field Resonance

- Instantaneous, non-local information transfer through coherence coupling, bypassing classical electromagnetic propagation.
- Emotional and cognitive communication across individuals via shared field domains — a scientifically grounded form of tele-empathy.

Biological and Medical Applications

- Healing by restoring field coherence — from cellular alignment to neural synchronization.
- Aging reinterpreted as coherence degradation, not time's erosion.
- Conscious intervention through meditation or field feedback as a measurable corrective process.

The laboratory will no longer be confined to instruments; the human field becomes part of the experiment.

"The next discovery won't come from smashing atoms, but from listening to alignment."

Toward a Coherence-Based Civilization

A coherence-based civilization is not one defined by technology, but by understanding. It values resonance over resistance, collaboration over competition, equilibrium over expansion. Its science serves consciousness, and its progress restores rather than consumes.

In such a civilization:

- Energy circulates, not depletes.
- Knowledge integrates, not divides.
- Progress aligns with nature, not against it.

Human creativity evolves into cosmic participation. We stop acting as isolated intelligences and begin functioning as coherent nodes of the universal field. In that realization, invention becomes remembrance — the rediscovery of what the universe has been doing all along.

"Civilization will mature the day it learns to design without disturbance."

"We are not here to conquer nature — we are here to complete her sentence."

Closing Reflection

The next paradigm of science is not new; it is ancient — as old as the first vibration. We are not creating coherence; we are finally learning to cooperate with it. Every experiment, every innovation, every act of awareness that aligns with balance brings us closer to the true renaissance — the moment when humanity stops measuring accidents and starts designing harmony.

"The future will not be built by force — it will be tuned by understanding."

"We once discovered phenomena by chance; soon we will design coherence by choice."

8.6 The Philosophical Fulfillment — Understanding as Alignment

The highest purpose of knowledge is not to expand information, but to reduce distortion. Every fact collected, every law discovered, every equation written — all are attempts to remove one more ripple of misunderstanding between the observer and the observed. In this sense, understanding is not accumulation — it is alignment.

Knowledge does not grow like a pile; it clears like a sky. The fewer the clouds, the brighter the light.

"To understand is not to add — it is to remove interference."

Knowledge as the Reduction of Distortion

For centuries, humanity has mistaken data for wisdom. The world became flooded with facts, yet clarity remained scarce. In truth, knowing more does not make one wiser; knowing clearly does. Every discovery in science, from Newton's gravity to quantum mechanics, was not a gain of information but a *correction* of perception.

UED redefines knowledge as ripple alignment within awareness — the state where observation ceases to distort what it observes. When understanding deepens, equations simplify, paradoxes dissolve, and separate disciplines converge naturally. It is not intelligence that expands; it is confusion that subsides.

True progress, therefore, is *clarification*, not complication.

"We call it discovery when distortion becomes too small to notice."

The Shift from Curiosity to Clarity

Curiosity is humanity's primal impulse — the search for what lies beyond. But curiosity without coherence is wandering; it multiplies questions faster than it resolves them. Curiosity led us to explore the atom, the cosmos, and consciousness — yet clarity remained elusive because the observer never stopped vibrating out of tune.

The next phase of human intelligence is the shift from curiosity to clarity — from reaching outward to resonating inward.

In this new paradigm, understanding is not an act of conquest but of consonance. It emerges when the observer and the observed vibrate in mutual coherence, when the question and the answer share the same field of stillness.

"Curiosity seeks; clarity resonates."

Understanding, then, is not discovery — it is resonance restored.

Understanding as Resonance

To understand something is to *vibrate with it correctly*. That is the secret all sages knew and all scientists eventually rediscover. When your thought aligns with the phenomenon, comprehension feels effortless — not because the brain has worked, but because the field has stopped resisting.

This is why great ideas arrive in silence, not in struggle. They appear whole, not assembled. They are not the products of thinking, but the consequences of coherence.

Every law, theory, or insight that stands the test of time resonates because it is *tuned* to universal alignment. False ideas eventually collapse — not because they are disproven, but because they fall out of phase with reality.

UED, therefore, defines understanding as sustained resonance between awareness and existence — a coherent feedback loop where knowledge becomes indistinguishable from being.

"Truth is what continues to resonate when the noise ends."

"Accidents End When Understanding Begins"

This is the heart of the UED Renaissance. The moment you understand the mechanism of coherence, accidents disappear — not because the universe becomes predictable, but because you finally see that it always was.

Accidents are illusions born from partial observation. Once coherence is recognized as the underlying constant, every "miracle," every "error," and every "random event" becomes intelligible. Understanding does not change the universe — it reveals that it never needed change.

"Accidents end when understanding begins, because misunderstanding was the only accident."

When cause and consequence are seen as one continuous motion, the mind stops fearing uncertainty.

To know the law is to trust the law; to trust it is to align with it; to align is to become it.

Awareness as Participation — The Scientist as Harmonizer

The age of the detached observer is over. The next scientist will not stand outside nature, recording her behavior; they will stand *within* her, tuning her coherence. Observation will evolve into participation, and participation into partnership.

The scientist of the future will not be a hunter of facts but a harmonizer of fields. The laboratory will become a listening chamber — a place where silence and symmetry meet. Measurement will no longer disturb, but refine. Knowledge will no longer divide, but unify.

This is the philosophical fulfillment of science — when understanding is no longer an act of knowing about the universe, but *knowing as* the universe.

"The true scientist is not the one who proves, but the one who listens until proof becomes unnecessary."

"When awareness participates, understanding completes itself."

Closing Reflection

Understanding, in its purest form, is the return of the observer to the observed — the completion of the circle that began with curiosity. It is the end of the human condition as accidental and the beginning of humanity as intentional. When knowledge aligns with coherence, truth ceases to be a pursuit and becomes a presence.

At that point, there are no mysteries left — not because everything is known, but because nothing is misunderstood.

"The mind that aligns with coherence understands everything — not by thinking more, but by being less distorted."

"Understanding is not achievement; it is remembrance of resonance."

8.7 The Renaissance of Understanding — Science Rediscovers Itself

Every civilization experiences a turning point — a moment when the language of ignorance can no longer hold the weight of understanding. The Middle Ages had its darkness; the Renaissance broke it not by force, but by illumination. Art, philosophy, and science began speaking again in harmony — each reminding the other that beauty, logic, and truth were never meant to be separate.

Today, humanity stands at the threshold of another renaissance — one not of invention, but of remembrance. For too long, science and spirit have stared at each other across a false divide. Now, through the lens of Universal Energy Dynamics (UED), they begin to meet again — as reflections of the same coherence.

"We were never inventing truth; we were remembering it in better language."

Parallels with the Historical Renaissance — From Dogma to Discovery

In the fifteenth century, the Renaissance liberated Europe from religious dogma through rediscovery — of geometry, anatomy, and cosmic motion. Humanity realized that understanding nature was not heresy but participation in divine order. Leonardo, Copernicus, Galileo — they didn't create truth; they unveiled alignment.

That first Renaissance was humanity's intellectual awakening — a recognition that observation and imagination belong together. But just as theology once restricted vision, modern science has done the same through its new dogma — mathematical absolutism.

Equations replaced scripture; probability replaced prophecy. We worshipped precision instead of perspective. Now, like the thinkers of Florence and Padua, the modern mind must rediscover the sacred — not in religion, but in continuity. The true Renaissance returns when science remembers that it was born from philosophy, not from competition.

"When philosophy lost patience, science was born. Now science must remember philosophy to grow wise."

The New Renaissance — Where Physics Returns to Philosophy

In the age of coherence, physics will no longer describe only the external — it will explain the internal logic of being. The universe will not be a collection of phenomena, but a continuum of intention. The next scientist will be both physicist and philosopher — fluent in mathematics yet mindful of meaning.

For too long, philosophy lacked evidence and physics lacked purpose. UED bridges the two by showing that intelligence and energy are the same process viewed from different scales. Philosophy describes what coherence feels like; physics describes how it functions. Together, they complete the circle of understanding — restoring wholeness to inquiry.

This synthesis marks the true Renaissance of Understanding — not a return to the past, but the rediscovery of what was never lost: that truth is both measurable and meaningful.

"The universe doesn't ask to be believed; it asks to be understood."

Education, Ethics, and Exploration — The Three Pillars of Coherence

Every renaissance transforms not only knowledge, but the way humanity learns, lives, and aspires. The Coherence Renaissance will bring the same transformation across three pillars:

Education — From Information to Illumination

- The classroom of the future will not produce specialists, but harmonizers.

- Students will learn not what to think, but how to align — seeing knowledge as field resonance rather than fragmented facts.
- Subjects will merge: physics with psychology, art with mathematics, biology with philosophy.
- The mind will no longer compete with memory — it will cooperate with understanding.

"Education is not filling the mind — it is tuning the instrument."
Ethics — From Commandment to Consequence

- Morality will no longer be dictated but realized — as coherence in action.
- The Law of Consequences (LOC) will redefine ethics as energy responsibility: every choice ripples through the shared field.
- The moral person will not obey rules; they will act in alignment because misalignment is energetically costly.

"Ethics is the science of alignment practiced as behavior."
Exploration — From Extraction to Participation

- Humanity's progress will no longer depend on conquest, but on calibration.
- Space travel will shift from thrust to resonance; biology from manipulation to cooperation.
- Exploration will mean extending coherence, not colonizing it.
- To explore responsibly is to participate in the universe's self-awareness.

"We don't conquer frontiers — we remember frequencies."

These three — education, ethics, exploration — are not separate disciplines, but reflections of one principle: the Law of Coherence. Together, they form the architecture of a civilization that understands rather than dominates.

UED — The Bridge Between Knowledge and Wisdom

The gap between knowledge and wisdom has always been the tragedy of human progress. Knowledge builds; wisdom balances. Knowledge accumulates; wisdom aligns. Knowledge sees parts; wisdom hears the pattern.

UED is the long-awaited bridge between the two — because it recognizes that wisdom is simply knowledge seen from the lens of coherence. It transforms complexity into simplicity, and theory into participation. It allows science to mature beyond mechanism and return to meaning.

Under the UED Renaissance, the purpose of knowing shifts from prediction to participation — from mastering laws to embodying them. When humanity learns to act *as coherence*, every discovery becomes an act of remembrance.

"Wisdom is knowledge that has found its alignment."

"We were never inventing truth; we were remembering it in better language."

Closing Reflection — Science Rediscovers Itself

Every renaissance is born from exhaustion — the exhaustion of error, of fragmentation, of noise. When knowledge stretches itself too thin, coherence calls it home. That moment has arrived again.

The coming age is not an age of machines, data, or algorithms — it is the Age of Coherence, where intelligence becomes self-aware of its own continuity. Science will rediscover what mystics always knew: that truth is not *out there* waiting to be found — it is *in here*, waiting to be aligned.

"The universe was never hiding truth — it was waiting for us to listen without distortion."

"The Renaissance returns not in paint or parchment, but in coherence — where knowing becomes remembering, and remembering becomes being."

8.8 Reflection and Closing — The End of Accidents

Humanity's journey through science has been, at its core, a journey through misunderstanding. We began by mapping consequences and mistook them for causes. We celebrated equations that worked, even when they didn't explain. We worshipped the measurable and ignored the meaningful. And so, the universe appeared random — not because it was, but because our instruments were tuned only to fragments.

But when the observer finally learns to listen instead of interfere, when awareness matures beyond curiosity into coherence — the illusion collapses. What once looked like probability becomes pattern. What once seemed chaos becomes conversation. And what we called "accidents" reveal themselves as unrecognized alignment.

"Accidents end when understanding begins."

From Quantum Confusion to Universal Clarity

We began this exploration with uncertainty — the quantum fog where matter blinked between states and observation appeared to decide reality. The probabilistic universe was born out of human limitation, not cosmic law. We built theories to explain our blindness and called them truth.

But Universal Energy Dynamics (UED) dissolves that fog. It restores clarity where confusion was institutionalized. It shows that beneath every quantum jump, every collapse, every statistical shadow lies continuous coherence — the quiet intelligence of the Structured Vacuum Energy (SVE) field.

Every phenomenon, from an electron's spin to a galaxy's spiral, follows the same principle: coherence seeking expression. There was never a dice-throwing deity; only an unbroken field resolving itself through lawful consequence.

"Nothing happens by chance — only by consequence."

This is not fatalism — it is freedom redefined. For when cause and consequence are understood as one continuum, choice itself becomes conscious participation in coherence. To act in awareness is to join the field's self-balancing dance. This is destiny, not in the mythic sense, but in the mechanics of understanding.

The Collapse of Probability into Coherence

Probability was humanity's confession — a mathematical apology for not knowing what connects events. It replaced comprehension with convenience. We called the unknown "random" simply because it escaped resolution in our frame of reference.

But UED reframes the so-called collapse not as destruction of potential, but as reconciliation of alignment. Every random event is a ripple returning to coherence. Every fluctuation is memory adjusting to balance. Every probability curve is the shadow of a single law — the Law of Consequence (LOC) — operating through scales of awareness.

Thus, randomness collapses not into certainty, but into clarity. The more we align our observation with the field's stillness, the less uncertainty we perceive. In the end, coherence replaces probability as the true foundation of order.

"The universe is not uncertain; it is undecoded."

Understanding as Destiny

From the first flicker of awareness to the last discovery of science, one motion has persisted — the motion toward coherence. That is the story of evolution, of thought, of civilization itself. Understanding is not a human invention; it is the universe remembering itself through us. Each realization, each breakthrough, each whisper of truth is a ripple returning home to silence.

This is the destiny of intelligence — not dominance, not conquest, but completion. To know is to align; to align is to fulfill the very purpose of existence.

Understanding is not the end of knowledge — it is its origin finally seen clearly. And in that moment, all dualities dissolve: observer and observed, cause and effect, human and universe. Only coherence remains — the still point that was always there, waiting for perception to settle enough to see it.

"We were never searching for truth; we were being searched by it."

The End of Accidents — The Beginning of Design

To end accidents is not to end spontaneity; it is to replace reaction with resonance. When coherence becomes conscious, every event gains meaning — not imposed, but inherent. The scientist, the artist, the philosopher, and the mystic all become participants in the same creative continuum.

This is the dawn of a new kind of civilization — one that designs harmony instead of documenting chaos. A civilization where science is no longer the study of fragments, but the architecture of alignment. Where progress is measured not by control, but by coherence. Where technology extends awareness rather than addiction.

"When humanity stops worshipping randomness, it will begin designing harmony."

The universe was never throwing dice — it was conducting a symphony. We simply mistook the tuning for noise. Now that we've learned to listen, the music can finally begin again — not as discovery, but as deliberate design.

Final Aphorisms

- "Nothing happens by chance — only by consequence."
- "Accidents end when understanding begins."
- "The universe is not uncertain; it is undecoded."
- "When humanity stops worshipping randomness, it will begin designing harmony."

Closing Reflection

The story of UED is not the end of physics — it is the return of philosophy to its rightful place within it. It does not destroy the old equations; it completes their intent. It shows that nature was never fragmented, only misread. The field was always whole; we only saw it through noisy minds.

And now, at the end of this journey — from matter to mind, from noise to silence, from chance to coherence — we arrive not at a conclusion, but at the beginning of awareness as participation. The universe was never an accident. It was, and remains, the art of alignment.

"To understand is to align. To align is to remember. To remember is to return."

"The end of accidents is the beginning of awareness."

Epilogue – The Nobel for Alignment

There was a time when humanity celebrated *accidents* as discoveries. When chance encounters with truth were hailed as genius, and serendipity became the unofficial law of progress. Every age of science, from Newton's falling apple to the 2025 Nobel, has glorified the narrative of the accidental — the romantic myth that understanding is stumbled upon, not earned through alignment.

But as the dust of centuries begins to settle, a deeper pattern emerges — one that reveals that what we called accidents were never random blessings of fortune. They were simply the field's patience — moments when alignment accidentally occurred, when the observer's mind fell silent enough for truth to reflect itself. Every great discovery was a coincidence only to the distracted; to the aligned, it was inevitable.

"When alignment becomes understanding, accidents stop winning Nobels."

1. The Age of Scientific Luck

The story of science is filled with legends of chance. Fleming leaves his Petri dish uncovered — penicillin is born. Roentgen plays with cathode rays — the X-ray appears. Einstein daydreams about chasing light — relativity dawns. And every time, the world declares, *"How lucky!"*

Yet beneath each tale of fortune lies the same hidden condition — clarity. The discovery never belonged to the accident; it belonged to the mind *ready enough* to see what the accident revealed. The world saw coincidence; nature saw coherence. Luck has always been the name we give to unrecognized alignment.

The Nobel Prizes of the twentieth and twenty-first centuries, though monumental, often rewarded the effect of coherence without honoring the conscious pursuit of it. They celebrated the finder, not the field; the result, not the resonance. Science advanced by accident — but every accident was a quiet echo of the same law: the Law of Consequence (LOC), which governs all realignment within the Structured Vacuum Energy (SVE) continuum.

The human mind was busy classifying outcomes while the universe was busy restoring balance.

2. From Discovery to Design

There comes a moment in any discipline when repeating accidents begins to feel inefficient. You can only stumble upon harmony so many times before you start wondering if there's a way to walk there intentionally. That is where humanity now stands — at the threshold between chance and choice, between accidental truth and conscious design.

The true future of science will not depend on the next particle found, but on the next alignment achieved. The laboratory of tomorrow will not be a place of trial and error, but of resonance and calibration. The experimenter will become a listener, not a breaker. Equations will evolve into expressions of rhythm — the measured poetry of coherence.

This shift marks a profound humility in the human project: to admit that we were never inventors of truth, only temporary translators of it. The great achievement of science is not that it uncovered nature's secrets, but that it finally realized there were no secrets — only unread patterns.

"We do not discover the universe; we decode it."

3. The New Meaning of Nobel

For more than a century, the Nobel Prize symbolized the pinnacle of human intellect — a recognition of discovery, insight, and contribution. Yet, in truth, it often rewarded those who arrived early to an inevitable consequence. The real prize was never for the idea itself, but for being in alignment with the moment it was ready to surface.

What would a Nobel look like in an age of coherence? It would no longer glorify separation — individual brilliance, isolated labs, competitive patents — but recognize *synchronization.* It would celebrate those who designed harmony, not those who merely observed it by chance. The Nobel for Alignment would honor conscious participation in coherence, not just its measurement.

It would belong not to the discoverer of a phenomenon, but to the harmonizer of fields — the one who helps understanding flow smoothly through humanity. Because in truth, science does not progress by discovery alone; it progresses by reduction of distortion. Every new equation, every new experiment, every new insight that brings clarity instead of confusion is an act of coherence design.

"The true scientist is not the one who finds what others missed — but the one who hears what the universe was already saying."

This new Nobel would not celebrate intelligence in isolation; it would recognize intelligence in alignment. It would see genius not as deviation from the norm, but as resonance with the natural order. It would measure contribution not by novelty, but by reduction of noise. It would be awarded not for *luck*, but for listening.

4. The Transition from Accident to Awareness

Every human discovery follows the same cycle: accident, interpretation, correction, and then humility. The first phase is the spark — a phenomenon noticed, an anomaly that refuses to behave. The second phase is the human story — we assign explanations, sometimes right, often premature. The third phase is the correction — deeper experiments reveal that what we thought was final was only partial. And the fourth phase — humility — rarely arrives, because by then we've already institutionalized the story.

UED marks the arrival of that fourth phase — the return of humility as method. It teaches that the field was never random, that the mind's chaos is not the universe's language. It shows that chance is a construct — an artifact of limited awareness. What we call "accidental discovery" is merely a delayed recognition of consistent consequence.

The human being who discovers coherence is not special — they are *aligned*. Their instrument — the mind — happened to fall into rhythm with the universal field. That moment of alignment produces what we call insight.

And insight, in UED terms, is not brilliance. It is clarity meeting opportunity.

"Every insight is the field whispering, 'Finally, you're quiet enough to listen.'"

The scientist who understands this will stop chasing accidents and begin designing alignment — both within themselves and within the systems they study.

5. When Science Becomes Self-Aware

There is a profound moment in every civilization when its tools begin to reflect its own mind. We are entering that moment now. Our telescopes peer deep into space, but they also show us the history of our own looking. Our colliders break matter, but also expose the breaking habit of thought. Our equations describe nature, but also reveal our preference for certainty over coherence.

Science has matured enough to ask not, *What is the universe made of?* but *What kind of intelligence does the universe express?* And in that question, the search for the divine and the search for the physical finally meet.

For UED, intelligence is not something *within* matter; it is the reason matter exists at all — the structured stillness that allows ripples to hold shape. Every law of physics is an expression of that intelligence maintaining balance. When we align our inquiry with that balance, we are no longer studying the universe — we are participating in it.

This is the dawn of self-aware science — where observation and observer are no longer separate, but entangled in shared coherence. The Nobel for Alignment would therefore not reward the one who observes best, but the one who *integrates most completely*.

"The future of science is not to observe more — it is to interfere less."

6. The Humility of the Field

The field has no ego. It doesn't reward or punish, discover or hide. It simply waits for alignment. Every ripple of energy, every pattern of matter, every thought in a conscious mind — all are allowed within its equilibrium. It does not choose favorites, yet it rewards harmony automatically.

The field's humility is its genius: it reflects whatever you bring to it. If you bring noise, it gives you interference. If you bring clarity, it gives you understanding. If you bring fear, it gives you resistance. If you bring awareness, it gives you revelation.

This is the cosmic impartiality that governs all discovery. The universe never grants truth to anyone; it simply becomes visible when alignment occurs. And that alignment is not an act of will — it is an act of listening.

To win the Nobel for Alignment, therefore, is not to stand above the world, but within it — attuned, receptive, humble. It is to realize that every discovery has already been waiting in silence, and that the mind's only role is to quiet itself enough to recognize it.

"The greatest minds are not inventors — they are listeners who forgot to interfere."

7. The Future of Recognition

Imagine a world where prizes are no longer given for invention, but for integration. Where recognition is not for creating new phenomena, but for reducing fragmentation. Where every field of knowledge — physics, medicine, psychology, art — converges around a single value: coherence.

The Nobel for Alignment would not be a medal, but a milestone — a symbol that humanity has matured from seeking truth to *living it.* It would honor the shift from accidental discovery to conscious participation in universal intelligence.

The recipients of this new Nobel would be the scientists who dissolve boundaries between disciplines. The educators who teach students how to *feel clarity* rather than memorize it. The engineers who design systems that sustain equilibrium instead of extraction. The philosophers who bring silence back into the language of inquiry. The artists who show that beauty and physics speak the same dialect — that symmetry and sincerity are one.

It would mark the collective awakening of a species no longer surprised by its own intelligence.

"When the Nobel is given for alignment, humanity will know it has finally stopped mistaking noise for progress."

8. The Era of Conscious Design

Designing coherence means consciously participating in the universe's natural order. It means recognizing that the same intelligence that organizes galaxies also organizes thoughts — that creativity, gravity, and morality are reflections of the same ripple logic.

When science begins to design coherence instead of disrupting it, it will step into its true role — not as a master of matter, but as a custodian of consequence.

Energy generation will shift from consumption to synchronization. Medicine will evolve from intervention to resonance. Education will move from memorization to understanding. Economy will transform from competition to contribution. And progress itself will be redefined — not as acceleration, but as alignment.

This is the destiny of intelligence — to mature from manipulation to meaning. To realize that control was never mastery, only misunderstanding. To replace the race for innovation with the rhythm of integration.

"The highest technology is not invention — it is alignment."

The Nobel for Alignment is not a fantasy — it is an inevitability. It is the natural next step in humanity's evolution from curiosity to coherence.

9. The End of Worship, the Beginning of Wisdom

For centuries, we have worshipped what we did not understand. The gods of myth, the particles of physics, the formulas of fate — all were placeholders for the same yearning: to find the law beneath the noise. But reverence without realization is only another form of ignorance.

When the field is understood, worship becomes wisdom. The sacred returns, not as faith, but as function. The divine no longer sits in temples or equations, but in equilibrium — everywhere coherence holds form.

The Nobel for Alignment, therefore, is not a prize for the elite — it is the moment when humanity itself earns recognition for remembering what it always was: a ripple within a coherent field, capable of awareness, creativity, and compassion.

"The divine was never outside us — it was the alignment we forgot to maintain."

10. Final Reflection — The Moment of Alignment

Imagine that one day, a Nobel ceremony unfolds in silence. No applause, no speeches, no competition. Only a shared recognition of coherence — of scientists, thinkers, and creators realizing they were all tuning the same field. A collective moment when humanity finally understands that truth was never discovered — it was remembered.

That will be the true Nobel — not awarded, but attained. Not a medal, but a milestone. Not for one person, but for the species that stopped confusing chance with consequence.

The future belongs not to those who discover more, but to those who disturb less. The greatest reward will not be fame, but resonance — the quiet satisfaction of being in tune with existence itself.

And in that silence, the universe will finally whisper back:

"Now you are listening."

Closing Aphorisms

- "Nothing happens by chance — only by consequence."
- "The universe never hides — it waits for coherence."
- "When alignment becomes understanding, accidents stop winning Nobels."
- "The future of discovery is design — the art of deliberate harmony."
- "When humanity learns to listen, silence will be its loudest breakthrough."

Final Line:

The Nobel for Alignment will not mark the end of science — it will mark the beginning of its wisdom. The moment when the universe no longer surprises us — because, finally, we are aligned enough to recognize ourselves within it.

"When alignment becomes understanding, accidents stop winning Nobels."

Final Author's Note

There are moments in life when knowledge stops feeling like accumulation and starts feeling like remembrance. Writing this book was one such moment for me. I did not set out to build a new theory, nor to challenge the old ones. I was simply trying to listen — to the silence that lies beneath equations, beneath arguments, beneath the noise of both faith and reason. What emerged was not invention. It was recognition.

Universal Energy Dynamics was never meant to compete with physics; it was born from its silence. It grew out of the spaces that traditional science had quietly avoided — the questions that felt philosophical, the paradoxes dismissed as poetic, the uncertainties labeled "unobservable." But truth, I have learned, does not hide behind complexity. It hides behind habit. We are trained to look for it in fragments — in particles, formulas, probabilities — while it waits patiently in continuity.

Over the years, as I taught, researched, and reflected, I realized something profoundly simple: *nature never needed us to explain it — only to understand it.* The universe is not a problem; it is a pattern. The field never breaks its law; we break our alignment with it. And when we return to that stillness — whether through science, meditation, art, or simple honesty — we find that the same intelligence that shapes galaxies also shapes our thoughts. Understanding, then, is not about decoding nature; it is about remembering we are part of it.

I have called this understanding *coherence* — the quiet law beneath all motion. To me, coherence is not just a scientific idea. It is an attitude toward existence. It is how you speak, how you think, how you work, how you love. It is what remains when interference ends. Every act of clarity — in science, art, or human kindness — is a moment of alignment. And in that alignment, life itself becomes intelligible.

There was a time when I too believed in the romance of discovery — that progress was built on accidents, on sudden sparks of genius or luck. But now I see that every so-called accident was simply the field aligning itself through us. The genius was not in the mind but in the moment. We were just fortunate enough to be silent when it spoke.

If I could leave one message behind, it would be this: *stop chasing certainty and start cultivating clarity.* Certainty is brittle; clarity is alive. Certainty divides knowledge into "known" and "unknown." Clarity unites them through understanding. And when you live in clarity, the universe no longer feels like a puzzle — it feels like a conversation.

For those who read this book as science, I hope it offers you reason. For those who read it as philosophy, I hope it offers you peace. For those who read it as faith, I hope it restores your wonder — without illusion. And for those who read it as curiosity, I hope it rewards you with silence — the kind that doesn't demand answers because it has already found rhythm.

This work is not finished, and perhaps it never should be. Because the moment truth feels final, it stops being alive. Each chapter here — from coherence to consciousness — is only a ripple within a greater field still unfolding. You and I, reader, are part of that unfolding. Every insight you gain adds to the universe's remembrance of itself.

We often say that we seek truth. But the truth, I think, seeks us — through patience, through consequence, through alignment. And when we stop running long enough, it finds us in silence. That silence is not emptiness; it is completion. It is the point where understanding becomes being.

So, if there is a legacy to this book, let it not be my name or my theory, but a reminder — that *the universe is listening*. And it responds to coherence.

Wherever you are reading this — in a classroom, a laboratory, or the quiet of your own mind — may you remember: the next great discovery is not out there; it is within your awareness. You will not find it by asking louder questions, but by becoming a quieter listener.

And when that moment of alignment arrives — that perfect stillness between curiosity and knowing — you will understand what I mean when I say:

"When alignment becomes understanding, accidents stop winning Nobels."

Appendices

"The universe never hides — it waits for coherence."

Appendix A – Timeline of Accidental Discoveries

Short Stories of Serendipity: X-rays, Penicillin, Cosmic Background Radiation, Superconductivity, and the 2025 Nobel

Science often advances not by the certainty of intent, but by the grace of coincidence. Throughout history, discovery has worn the mask of accident — yet every so-called accident was a ripple aligning with coherence before human awareness could catch up. The following timeline traces some of those moments when nature briefly opened its veil — not to reward curiosity, but to remind us that she has always been coherent, waiting for us to notice.

1895 — X-Rays: The Light That Passed Through Everything

Wilhelm **Conrad Röntgen** was not searching for invisible light. He was investigating cathode rays when his darkened lab flickered unexpectedly — a glow from a barium screen several feet away. Something unseen had passed through solid matter, leaving an imprint on photographic plates.

He named them *X-rays*, because "X" was the symbol of the unknown. The world celebrated an accident; medicine celebrated a miracle. But under UED, this was not a random gift — it was coherence breaking through interference.

Röntgen's setup had unintentionally created a low-noise condition: vacuum, voltage, and shielding formed a temporary alignment where ripple energy propagated without obstruction. The universe wasn't hiding X-rays; it was waiting for alignment precise enough to let the silent frequency of penetration become visible.

"Discovery is what stillness looks like when the noise of ignorance turns off."

1928 — Penicillin: The Mold That Listened

Alexander Fleming left his Petri dishes uncovered during a summer holiday. Upon returning, he found bacteria had refused to grow around a patch of mold. From this oversight came *Penicillin* — a medicine that transformed civilization.

Science called it luck; medicine called it salvation. But under UED, even life's smallest organisms obey coherence. The mold and bacteria were not fighting — they were seeking equilibrium. Fleming's accidental contamination allowed one field of life to restore balance against another — a natural correction of microbial misalignment.

What looked like chaos was nature rehearsing order. When environmental interference (temperature, humidity, contamination) crossed a precise threshold, the field reorganized itself — favoring survival through balance. Fleming merely witnessed the ripple realigning life's tension through consequence.

"Every cure begins as conflict — until the field remembers balance."

1964 — Cosmic Background Radiation: The Sound of the Universe Remembering

Two radio engineers, **Arno Penzias and Robert Wilson**, were not hunting for the universe's origin. They were trying to reduce a persistent noise in their antenna — an irritating hiss that refused to go away. They cleaned, recalibrated, chased pigeons off the dish —

yet the noise persisted. Eventually, they realized it wasn't noise at all — it was the residual whisper of creation itself.

That faint microwave radiation, uniform across the sky, was the first empirical echo of the Big Bang — the cosmic background field.

In UED interpretation, this discovery was far from accidental. Penzias and Wilson had unknowingly aligned their instrument with the baseline ripple frequency of the Structured Vacuum Energy (SVE). The so-called "background" was not a leftover — it was the continuing hum of equilibrium, the eternal field breathing beneath all change.

The noise they tried to remove was coherence itself, misunderstood. They didn't eliminate interference; they learned to listen through it.

"Sometimes, noise is the universe speaking in a frequency we forgot to understand."

1911 — Superconductivity: The Cold Realization of Order

Heike Kamerlingh Onnes, experimenting with metals cooled near absolute zero, expected resistance to decrease — but not disappear. When he observed mercury carrying current *without any measurable resistance*, he realized something unprecedented: the material had entered a state of perfect order.

No collisions. No loss. Just pure flow. Superconductivity had been discovered — though no one knew why.

From the UED perspective, this was not an anomaly — it was nature's reminder that stillness perfects function. Cooling did not create a new behavior; it removed misalignment. When thermal agitation fell below the coherence threshold, ripple tension unified across the conductor. The current did not move; it resonated. The field didn't conduct energy — it became energy in alignment.

Onnes thought he found a new phenomenon. In truth, he had just accidentally silenced interference long enough to hear what matter had always been saying.

"Stillness doesn't stop motion — it perfects it."

2025 — The Nobel for Macroscopic Coherence: The Tunnel That Wasn't There

John Martinis, John Clarke, and **Michel Devoret** were awarded the 2025 Nobel Prize in Physics for achieving macroscopic quantum coherence in superconducting circuits — a feat that seemed to prove that quantum behavior could exist at human scale.

They called it *tunneling* — the ability of energy to cross barriers without sufficient cause. The world saw wonder; physicists saw paradox.

But under Universal Energy Dynamics, there was no paradox at all. No particle "passed through." No wall was breached. The so-called tunneling was simply the field remembering its coherence.

Cooling, shielding, and vibration isolation had created a domain so quiet that ripple tension could reorganize continuously across the circuit. The barrier didn't vanish — it *equalized.* The energy didn't move — it *resolved.*

Their discovery was not a quantum miracle; it was nature revealing stillness through human patience.

"They didn't break through; they aligned through."

In celebrating this experiment, humanity once again mistook silence for surprise. Yet this "accident" — like all before it — was coherence gently asserting itself through the cracks of probability.

The Pattern Behind Every Accident

Across centuries, from Röntgen's rays to Devoret's qubits, the story repeats: disorder gives birth to discovery only because order was always present, waiting. Each so-called serendipity is not a gift of chance but a temporary synchronization between human intent and cosmic rhythm.

Every discovery requires two conditions:

1. Nature's readiness to reveal, and
2. Human willingness to listen.

Accidents occur when both align unconsciously; understanding occurs when the alignment becomes conscious. The moment humanity realizes this, science itself transforms — from a game of probability into a practice of coherence.

"The accident ends the moment you understand why it was inevitable."

The Unbroken Thread of Alignment

Let us step back and look at the continuum:

Year	Discovery	Conventional Interpretation	UED Translation
1895	X-Rays (Röntgen)	Random emission of invisible light	Ripple field coherence penetrating dense media through perfect alignment
1911	Superconductivity (Onnes)	Electrical resistance vanishes mysteriously at low temperature	Ripple tension achieves full alignment; motion becomes resonance
1928	Penicillin (Fleming)	Mold contamination kills bacteria accidentally	Biological fields rebalance misalignment naturally; coherence restores harmony
1964	Cosmic Background Radiation (Penzias & Wilson)	Random leftover signal from Big Bang	Structured Vacuum Energy hum — baseline coherence of the universe
2025	Macroscopic Coherence (Martinis, Clarke, Devoret)	Quantum tunneling extended to large scale	Ripple field realignment through equilibrium — no tunneling, only stillness

Each milestone tells the same story in different dialects — the story of coherence rediscovered through accident.

The Moral of Serendipity

If every "accident" has a pattern, it ceases to be an accident. It becomes a language. A dialogue between curiosity and consequence.

The field is not random — it is responsive. When we act with clarity, nature mirrors that coherence back. When we act with confusion, it reflects noise. Serendipity, then, is not luck; it is the field's way of teaching alignment through experience.

"Serendipity is the universe's polite reminder that it was never lost — only unheard."

Epilogue Within the Appendix

Looking across this timeline, it becomes clear that the greatest discoveries did not arise from control — they arose from carelessness that became clarity. From openness that allowed interference to fall away. The accidents of science are not random cracks in reality; they are points of contact between human perception and universal coherence.

Each discovery was a mirror moment: Röntgen saw through matter, Fleming through life, Penzias through noise, Onnes through stillness, and Martinis through illusion. Each lifted a veil — but none tore it.

Because the veil was never between us and truth; it was between *noise and listening.*

And perhaps, when humanity learns to align without stumbling — when experiments are guided by resonance rather than probability — there will be no more accidental discoveries. There will only be understanding unfolding at the natural pace of coherence.

"When understanding aligns with consequence, serendipity retires — and science becomes remembrance."

Appendix B – UED Glossary for Readers

A field guide to the language of coherence

Universal Energy Dynamics (UED) uses a language that merges physical precision with philosophical insight. The words below are not technical inventions but translations—bridges between science's measurable terms and the universe's deeper intent. Each concept expresses a different aspect of one truth: reality is structured stillness in motion.

1. Ripple

Definition: A localized disturbance in the Structured Vacuum Energy (SVE) field that represents the dynamic expression of equilibrium seeking to restore itself.

In simple terms: A ripple is existence in action—the universe remembering itself through motion.

Expanded Meaning: Every observable phenomenon—light, matter, sound, thought—is a ripple, not an object. A ripple does not travel *through* space; it reshapes the field that contains it. It is both event and environment, cause and consequence. In UED, ripples are not separate entities but temporary coherences, born when the field's stillness distorts slightly and seeks alignment again.

Philosophical Reflection: "We are not made of matter; we are made of memory—ripples that remember stillness."

2. SVE (Structured Vacuum Energy)

Definition: The foundational substrate of existence—a continuous, tension-bearing medium of ordered potential that underlies all phenomena.

In simple terms: SVE is the silent ocean from which all waves emerge.

Expanded Meaning: Conventional physics calls it vacuum; UED calls it structure. It is not emptiness but equilibrium—an active stillness possessing latent tension, capable of sustaining ripples without fragmenting. When disturbed, it generates observable energy, form, and consciousness. When undisturbed, it remains the zero-point of intelligence—the unbroken continuity behind time and matter.

Analogy: If the physical world is music, SVE is the instrument that never stops resonating.

Philosophical Reflection: "The vacuum is not nothing—it is everything that has not yet forgotten silence."

3. ATT (Alignment Threshold Time)

Definition: The minimal temporal window required for a ripple or system to realign with the field's equilibrium.

In simple terms: ATT is how long stillness takes to remember itself.

Expanded Meaning: Every system—atomic, biological, mental—has a characteristic ATT, a moment when coherence becomes possible. If alignment is attempted too early, turbulence persists; if sustained beyond ATT, the system stabilizes. It defines the rhythm of natural processes—from the recovery time of neurons to the formation of galaxies.

Scientific Implication: ATT links microscopic and cosmic behavior through timing of alignment, providing a measurable bridge between quantum relaxation, thermodynamic equilibrium, and biological regeneration.

Philosophical Reflection: "Patience is physics in disguise—the field realigns only when the moment is ready."

4. LOC (Law of Consequences)

Definition: The universal principle governing all ripple interactions: every disturbance seeks equilibrium through coherent redistribution of tension.

In simple terms: LOC is the law of balance acting as destiny.

Expanded Meaning: Unlike Newton's action–reaction, LOC includes intelligence—the inherent capacity of the field to self-correct. Energy never disappears; it redistributes until coherence is restored. All physical, mental, and moral events follow LOC. In human terms, it is Karma translated into field mechanics.

Mathematical Form (Conceptual): $\Delta T \to 0 \Rightarrow \Delta C \to \infty$ When tension variation approaches zero, coherence tends toward infinity.

Philosophical Reflection: "Nothing happens by chance—only by consequence seeking rest."

5. Ripple Tension

Definition: The intrinsic elastic strain within the SVE field that enables ripples to exist, interact, and sustain form.

In simple terms: Ripple Tension is the field's stored intent—energy held as the possibility of movement.

Expanded Meaning: It corresponds to what classical science measures as force, charge, or pressure, but it is not transmitted; it is redistributed. Every particle's mass, every magnetic line, every emotional impulse is a manifestation of ripple tension adjusting within equilibrium. When tension is uniform, the field appears still; when uneven, phenomena arise.

Scientific Analogy: In superconductivity, when ripple tension aligns perfectly, resistance disappears—the field no longer fights itself.

Philosophical Reflection: "All force is borrowed tension; only stillness owns itself."

6. Ripple Coherence

Definition: The degree of alignment among ripple phases within a given region of the SVE field.

In simple terms: Coherence is the harmony of existence.

Expanded Meaning: When ripples oscillate in mutual phase, energy flows effortlessly; when out of phase, resistance, friction, and heat emerge. Coherence determines the effectiveness of communication, stability of matter, and even the quality of thought. Superconductivity, intuition, empathy, and cosmic order are all scales of the same coherence principle.

Scientific Analogy: Laser light is coherent electromagnetic ripple; enlightened awareness is coherent cognitive ripple. Both are clarity achieved through alignment.

Philosophical Reflection: "Coherence is the intelligence of the field remembering its rhythm."

7. Structured Stillness

Definition: The ultimate equilibrium state of the SVE field—pure order without motion, yet containing the potential for all motion.

In simple terms: Structured Stillness is silence made intelligent.

Expanded Meaning: It is not emptiness but perfect alignment—the universal baseline from which every ripple arises and to which all return. Every phenomenon, from thought to thunder, is a temporary deviation from this state. Structured Stillness is the true meaning of "zero energy"—not nothingness, but infinite coherence without contrast. In spiritual terms, it corresponds to consciousness without object; in physical terms, to field without tension.

Philosophical Reflection: "Stillness is not the absence of life—it is life before the need to move."

Unified View: The Ripple Chain of Reality

Concept	Role in UED	Analogy
SVE	The silent substrate—the still ocean.	Vacuum as structure, not emptiness.
Ripple Tension	The local strain that allows expression.	The string drawn before vibration.
Ripple	The event of movement born from imbalance.	The wave upon the ocean.
Ripple Coherence	The rhythm that makes waves harmonize.	Music from coordinated vibration.
ATT	The time required for the wave to retune.	Rest between heartbeats.
LOC	The law that restores harmony after deviation.	Karma of physics.
Structured Stillness	The final equilibrium—the silent completion.	Ocean without waves.

Together they form the Seven Pillars of UED Language, describing how energy, intelligence, and existence are not separate things but different moods of one still field.

Closing Reflection

Language, at best, can only gesture toward what UED describes. The field does not speak in words but in resonance. These terms—Ripple, SVE, ATT, LOC, Ripple Tension, Ripple Coherence, Structured Stillness—are stepping stones, not conclusions. They allow the intellect to touch what intuition already knows: that beneath every phenomenon lies a continuous intelligence maintaining balance.

When you understand these seven words not as definitions but as states of alignment, the field ceases to be theory—it becomes presence. You begin to sense it in conversation, in silence, in breath, in thought. And then, the glossary is no longer a guide; it becomes a mirror.

"Every definition is a ripple; only understanding is stillness."

Appendix C – The Law of Consequences (LOC)

The *Law of Consequences (LOC)* is the foundational dynamic of Universal Energy Dynamics (UED).

It states that:

Every disturbance within the Structured Vacuum Energy (SVE) field seeks to restore equilibrium through coherent redistribution of tension.

In this view, *cause and effect* are not linear events but simultaneous expressions of one self-correcting field.

1. Symbolic Form

Let

- T = Ripple Tension (localized deviation from equilibrium)
- C = Coherence (degree of alignment)
- E = Energy State of Field
- Φ = Field Potential of SVE

Then the governing relation is:

$$\frac{dT}{dt} = -k\,C \quad \text{and} \quad \frac{dC}{dt} = +k\,T$$

where k is a proportionality constant expressing the field's tendency toward re-alignment. At perfect equilibrium:

$$\Delta T \to 0 \Rightarrow \Delta C \to \infty$$

meaning that as tension vanishes, coherence becomes complete—Structured Stillness.

2. Interpretation

Aspect	Conventional View	UED Translation
Cause → Effect	Sequential action	Simultaneous realignment
Entropy	Irreversible decay	Local misalignment increasing
Energy Loss	Dissipation	Redistribution of tension
Equilibrium	Static balance	Dynamic coherence
Karma (Philosophical)	Moral reaction	Field correction

3. Operational Cycle of LOC

1. **Disturbance** – a misalignment or ripple arises within the SVE field.
2. **Differentiation** – tension creates local asymmetry; phenomena become observable.
3. **Interaction** – ripples exchange tension through energy transfer, appearing as motion or reaction.
4. **Re-alignment** – coherence restores equilibrium; tension dissolves.
5. **Memory Imprint** – the field retains structural intelligence of the event, refining future balance.

"Consequence is memory practicing correction."

4. Domains of Application

- Physical – forces, thermodynamics, electromagnetism, gravitation.
- Biological – cellular repair, neural adaptation, ecological balance.
- Psychological – emotion regulation, thought alignment, moral awareness.
- Cosmic – galactic structure, cosmic expansion, black-hole equilibrium.

In every domain, LOC operates as the unseen algorithm of restoration.

5. Graphical Representation (Conceptual)

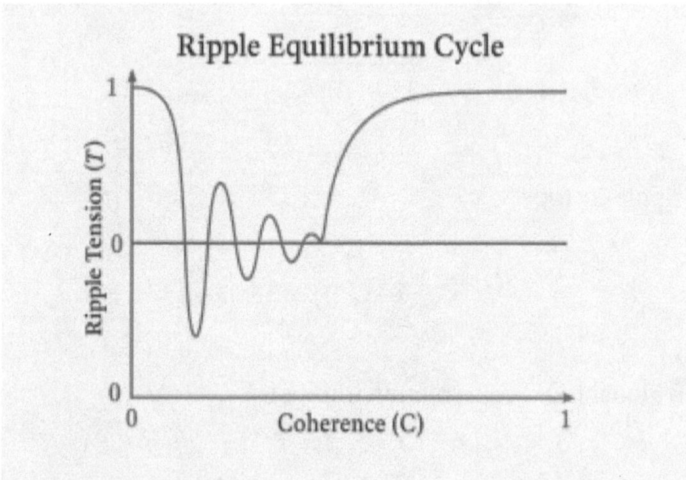

Interpretation: each cycle of tension and release refines equilibrium; the amplitude diminishes with understanding—intelligence learning its own rhythm.

6. Philosophical Implications

- Causality Reimagined: LOC replaces mechanical cause-effect with *interactive consequence.*
- Morality as Physics: ethical action is coherence-preserving behavior; unethical acts are misalignments generating compensatory tension.
- Time as Correction: temporal flow is simply the sequence of realignments perceived by consciousness.
- Evolution as Equilibrium Practice: life evolves by refining alignment between local and universal fields.

"The universe does not punish; it balances."

7. Mathematical Heuristic (Extended)

Define coherence density ρC as:

$$\rho_C = \frac{1}{1 + \rho_{misalignment}}$$

and ripple entropy S_r as:

$$S_r = k \ln(\rho_{misalignment})$$

Then global LOC conservation implies:

$$E_{total} = \rho_C + S_r = \text{constant}$$

Thus, loss in coherence locally equals gain in ripple entropy elsewhere—no destruction, only redistribution.

8. Ethical Corollary

When applied to human behavior:

$$\text{Intent aligned with field} \rightarrow \text{Constructive Consequence}$$

$$\text{Intent misaligned} \rightarrow \text{Corrective Consequence}$$

The LOC therefore unites physics and ethics under one principle: *"Energy does not judge you; it corrects you."*

9. Ultimate Formulation

$$\boxed{\Delta T + \Delta C = 0}$$

Every increase in coherence neutralizes equal tension elsewhere. The universe's total equilibrium never changes—it simply learns new ways of maintaining it.

Closing Reflection

The Law of Consequences is not a rule among many—it is the rhythm of existence itself. From atomic spin to human choice, from gravitational curvature to emotional release, every act of creation and correction follows the same pulse: misalignment → resolution → memory.

Science describes it as energy conservation. Philosophy calls it karma. UED calls it coherence remembering itself.

"The universe does not react; it resolves."

Appendix D – The Law of Ripple Coherence

Mathematical Form, Experimental Implications, and Falsifiable Predictions

1. Conceptual Foundation

The **Law of Ripple Coherence** (**LRC**) governs how localized disturbances in the Structured Vacuum Energy (SVE) field regain equilibrium through phase synchronization. It expresses the universal tendency of all ripples—whether in matter, light, thought, or gravity—to seek *coherent alignment*.

If the *Law of Consequences (LOC)* defines the *why* of restoration, then the *Law of Ripple Coherence (LRC)* defines the *how*.

In essence:

"Every ripple seeks to resolve itself through coherence — not motion."

2. Mathematical Formulation (Symbolic Representation)

Let:

- $\psi(x, t)$ = Ripple function of local field disturbance.
- $\phi(x, t)$ = Phase potential of underlying SVE at that region.
- $C(x, t)$ = Coherence density ($0 \leq C \leq 1$).

- $\rho_m(x, t)$ = Misalignment density.
- τ = Relaxation time constant of alignment (analogous to ATT).
- η = Ripple tension coefficient (local elasticity of the field).

Then the dynamic relation of coherence evolution is:

$$\frac{\partial C}{\partial t} = \frac{1}{\tau}(1 - C) - \eta \nabla^2 C$$

This expresses that coherence increases with time until interference gradients ($\nabla^2 C$) dissipate. Equilibrium is reached when $\partial C/\partial t \to 0$, implying a self-sustained alignment pattern—what UED calls Structured Stillness.

Complementary Formulation:

$$\rho_m = (1 - C)$$

$$E_{local} = \eta(\rho_m)^2$$

→ Misalignment produces stored tension energy, released upon re-coherence.

Thus, the system's energy does not vanish—it redistributes as phase realignment.

3. Interpretation

Term	Meaning under Classical Physics	Reinterpretation under UED
ψ(x,t)	Wavefunction / probability amplitude	Ripple configuration of field tension
C(x,t)	Coherence probability	Phase harmony density
ρ_m	Disorder / entropy	Misalignment intensity
τ	Decay constant	Alignment Threshold Time (ATT)
η	Resistance or damping	Ripple stiffness / field elasticity

Under this model, wave behavior, temperature gradients, and even neural synchronization become measurable manifestations of coherence flow rather than energy transport.

4. Experimental Implications

a. Superconductivity and Superfluidity

- Measure coherence function *C(x,t)* in superconducting materials during transition through critical temperature.
- Prediction: *C rises exponentially toward 1 as ripple tension ($\eta \rho_m{}^2$) approaches zero*, matching zero resistance onset.
- Testable via ultrafast coherence spectroscopy or Josephson phase tracking.

b. Optical Coherence in Nonlinear Media

- Coherence peaks (C ≈ 1) should correspond to nonlinear phase-locking even without photon quantization.
- UED predicts that "quantum" squeezing or entanglement effects will appear whenever $\partial C/\partial t \to 0$.

c. Biological Coherence

- Neural networks or heart-brain systems display rising C during meditative or flow states.
- Detectable via EEG phase synchrony or HRV coherence.
- The model predicts a measurable ATT specific to each biological subsystem — confirming the "Law of Alignment Time."

d. Vacuum Resonance Experiment (Direct Test of UED)

- Setup: Two separated cavities with synchronized electromagnetic ripples.
- Prediction: Phase coherence transfer occurs *without energy exchange* when C > Ccritical.
- Detectable as correlated phase fluctuation reduction between independent systems — falsifying the photon-based transmission model.

5. Falsifiable Predictions

Prediction	Testable Outcome	Methodology
1. Non-quantized coherence transfer	Energy phase-locks across barriers without discrete quanta.	Interferometry across superconducting junctions at sub-quantum limits.
2. Threshold of spontaneous coherence (Ccritical)	Below specific misalignment density, coherence self-propagates.	Low-temp critical domain studies.
3. Biological ATT	Predictable delay between noise suppression and coherence spike.	EEG + HRV cross-correlation.
4. Field Re-synchronization Memory	System returns to prior coherent state after disturbance without loss.	Repeated decoherence–recoherence cycles.
5. No photon drift in tunneling equivalents	Energy redistribution without particle motion.	Barrier-embedded field phase mapping.

Each prediction is falsifiable through measurable physical or biological experiments — confirming or refuting the coherence-first interpretation.

6. Relationship to Existing Laws

Existing Law	Reinterpreted by LRC	UED View
Conservation of Energy	Tension redistribution	No energy lost; only misalignment corrected.
Second Law of Thermodynamics	Local entropy rise, global coherence conserved	Disorder = localized misalignment.
Maxwell's Equations	Field propagation of ripples	Phase continuity, not charge transport.
Schrödinger Equation	Time evolution of wavefunction	Ripple alignment dynamic with built-in coherence term.

Thus, the LRC is the bridge uniting the mechanics of energy and the intelligence of equilibrium.

7. Visualization Concept

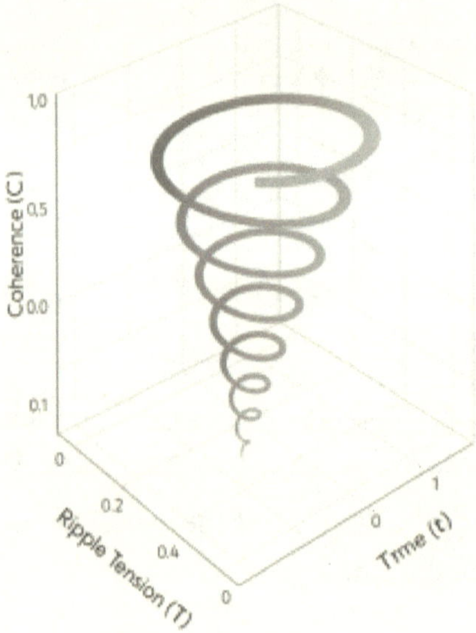

3D plot:

- X-axis → time (t)
- Y-axis → coherence (C) rising toward 1
- Z-axis → ripple tension (T) decaying toward 0 The trajectory spirals upward, showing that each cycle of tension release increases stable coherence—a learning curve of the universe itself.

8. Philosophical Reflection

Every scientific law describes *behavior*; the Law of Ripple Coherence describes *remembering.*

Matter remembers stillness. Life remembers intelligence. Mind remembers silence.

The field does not progress—it recalls its own equilibrium across scales. In this memory lies the blueprint of consciousness, creativity, and cosmic order.

"Coherence is the field's way of thinking — alignment is its language."

9. Closing Statement

The *Law of Ripple Coherence* transforms physics from a study of particles to a study of participation. It provides measurable, falsifiable principles while restoring continuity between science and sense.

Where probability once reigned, coherence takes its place. Where energy was said to move, tension now realigns. And where randomness ruled, understanding begins.

"The field doesn't seek perfection—it seeks participation in its own balance."

Appendix D – Quote Bank

Selected Aphorisms from Each Chapter for Public Use and Teaching

The Age of Lucky Discoveries

- "Every time humanity stumbled, nature took a step forward."
- "Accidents are coherence waiting to be recognized."
- "Progress is not invention — it's rediscovery in slow motion."
- "When understanding sleeps, luck wears the crown."

The Quantum Confession: The Birth of Uncertainty and the Death of Continuity

- "When math loses meaning, nature hides behind probability."
- "The first Nobel for confusion was called quantization."
- "Randomness is not nature's nature — it's our blindness in motion."
- "What we call uncertainty is coherence too fine for our instruments."
- "The cat never split — only our logic did."

The Tunnel That Wasn't There: The 2025 Nobel and the Myth of Quantum Tunneling

- "They didn't break through; they aligned through."
- "Every time coherence is achieved, humanity names it a miracle."
- "Tunneling isn't motion — it's misunderstanding."
- "Stillness doesn't pass through barriers; it makes barriers irrelevant."
- "They silenced the noise — and the field spoke."

The Cold Truth: When Matter Remembers Itself

- "At absolute zero, nothing freezes — everything listens."
- "The colder the matter, the clearer its memory."
- "Stillness reveals more than motion ever could."
- "Heat is not energy — it's misunderstanding made physical."
- "Zero is not the end of motion; it is the beginning of meaning."

The Accidental Symmetry: From Spin to Domains

- "Symmetry is nature's apology for earlier turbulence."
- "Magnetism is not attraction — it's the echo of shared stillness."
- "Order is not forced — it emerges when interference stops."
- "Every pattern of order is a memory of silence made visible."
- "Alignment is the invisible hand behind every symmetry."

The Noise and the Silence: Decoherence, Dissipation, and the Myth of Loss

- "Nothing disappears — it only forgets how to stay together."
- "Entropy is not decay — it's unremembered alignment."
- "The field never forgets; it only distributes memory too widely for you to find."
- "Silence is not the absence of sound — it's the sound that remembers."
- "Noise is coherence trying to come home."

The God Particle and Other Myths: What We Found When We Forgot What We Were Looking For

- "When you stop understanding, you start worshipping."
- "We named it divine because we couldn't describe it mechanical."
- "You can't find truth by shattering harmony."
- "Mass is stillness condensed — the memory of motion."
- "Spirit is not beyond matter — it's matter that remembered silence."

The Accidental Human: The Brain, the Field, and the Flash of Alignment

- "Ideas are not born — they occur when the field clears its throat."
- "Stillness is not the absence of thought — it is the mind's resonance with the field."
- "Emotion is energy remembering its shape."
- "Intuition is the mind's shortcut to alignment."
- "We are accidents only to those who don't understand

equilibrium."

The Universe Without Accidents: From Probability to Coherence

- "Every accident is order seen too early."
- "Accidents end when understanding begins."
- "The universe is not uncertain; it is undecoded."
- "There was never one equation — there was always one intention."
- "When humanity stops worshipping randomness, it will begin designing harmony."

The Nobel for Alignment

- "When alignment becomes understanding, accidents stop winning Nobels."
- "The universe doesn't reward luck; it rewards listening."
- "The next Nobel will not be for discovery — it will be for coherence."
- "Truth was never hidden — it was always waiting for silence."
- "Understanding is the final symmetry."

Teaching Notes

- These aphorisms can be used in lectures, classroom discussions, and public science outreach to bridge physics and philosophy.
- Each statement condenses a full argument into one reflective insight — designed for thinkers, educators, and students of coherence science.
- Readers are encouraged to treat them not as doctrines but

as lenses — ways to *see the field as conversation, not competition.*

Further Reading and References

For readers seeking deeper exploration of coherence, field theory, and the philosophical reconstruction of physics.

1. Classical Foundations of Modern Physics

- **Max Planck**, *On the Law of Distribution of Energy in the Normal Spectrum* (1900) — the origin of quantization and the "first accident."
- **Albert Einstein**, *Relativity: The Special and General Theory* (1916) — the great reconciliation between motion and curvature.
- **Erwin Schrödinger**, *What Is Life?* (1944) — where physics first glimpsed biology's alignment principle.
- **Werner Heisenberg**, *Physics and Philosophy* (1958) — the earliest admission that uncertainty was conceptual, not cosmic.
- **Richard Feynman**, *The Character of Physical Law* (1965) — a lyrical account of nature's hidden simplicity.

2. Quantum Mechanics and Its Dilemmas

- **Niels Bohr**, *Atomic Theory and the Description of Nature* (1934) — the philosophical paradox of observation and reality.

- **David Bohm**, *Wholeness and the Implicate Order* (1980) — the first serious attempt at a unified field of consciousness and matter.
- **John Bell**, *Speakable and Unspeakable in Quantum Mechanics* (1987) — the mathematics of non-locality that re-opened the field.
- **Carlo Rovelli**, *Helgoland* (2021) — relational quantum mechanics and the return of interpretation to physics.

3. Coherence, Symmetry, and the Physics of Alignment

- **H. Kamerlingh Onnes**, *Further Experiments with Liquid Helium* (1911) — discovery of superconductivity and the meaning of stillness.
- **Lev Landau**, *The Theory of Superfluidity* (1941) — the mathematics of frictionless existence.
- **John Bardeen, Leon Cooper, and Robert Schrieffer**, *Microscopic Theory of Superconductivity* (1957).
- **John Martinis, John Clarke, and Michel Devoret**, Nobel Prize Lectures (2025) — macroscopic coherence at the human scale.
- **David Deutsch**, *The Beginning of Infinity* (2011) — reflections on explanation as the essence of progress.

4. Bridging Physics and Philosophy

- **Werner Heisenberg**, *The Part and the Whole* (1969) — conversations revealing science's loss of metaphysical balance.
- **Thomas Kuhn**, *The Structure of Scientific Revolutions* (1962) — how paradigms collapse under their own noise.

- **Arthur Eddington**, *The Nature of the Physical World* (1928) — the earliest fusion of observation and ontology.
- **Rupert Sheldrake**, *Morphic Resonance* (1981) — speculative yet resonant with the UED concept of field memory.

5. Contemporary Field Theories and Unified Approaches

- **Lee Smolin**, *Three Roads to Quantum Gravity* (2001) — the struggle to merge geometry with quantum behavior.
- **Frank Wilczek**, *A Beautiful Question* (2015) — symmetry as aesthetic coherence.
- **Sean Carroll**, *The Big Picture* (2016) — the narrative of meaning within a mechanical cosmos.
- **Sabine Hossenfelder**, *Lost in Math* (2018) — the critique of abstraction divorced from intuition.

6. Works by Sandeep Chavan and the Universal Energy Dynamics (UED) Canon

For readers seeking the complete conceptual bridge between classical physics and structured stillness, the following companion works expand the framework developed in *The Nobel for Accidents*:

- *The Last Quantum Book* — resolving quantum paradoxes through ripple-field continuity.
- *Fields: The Invisible Architecture of Coherence* — understanding energy as structured stillness.
- *Ripple Logic of Electricity* — reinterpreting current, charge, and magnetism as field reorientation.
- *The Human Machine* — biological coherence and consciousness as field behavior.

- *The Divine Dilemma* (from *The War of Thoughts* series) — dialogue between ancient insight and modern physics.
- *Vedanta Reimagined: Beyond Myth, Toward Universal Science* — philosophical roots of UED in timeless awareness.

7. Suggested Exploration Themes

- **Field Listening Experiments** — studying coherence spikes in biological and neural systems.
- **Ripple Dynamics Simulations** — modeling phase alignment in superconductive circuits.
- **Entropy Reversal Research** — examining local restoration of order through controlled cooling.
- **Educational Frameworks** — integrating ethics of coherence into scientific pedagogy.

8. Closing Reflection

"The greatest experiment is not conducted in a laboratory — it is conducted in understanding.

When curiosity becomes coherence, science fulfills its promise."

Acknowledgements

This book was not born in isolation; it emerged through a long conversation — between thought and silence, reason and wonder, science and self. Every page owes its existence to that dialogue.

First, I offer my deepest gratitude to *the thinkers who dared to doubt* — Planck, Einstein, Schrödinger, Bohr, and all who stood at the edge of understanding. Their courage to question became the raw material of this reconstruction. Without their missteps, there would be no rediscovery.

To the *modern pioneers* — John Martinis, John Clarke, and Michel Devoret — whose 2025 Nobel work rekindled this reflection. Your experiments revealed not just quantum coherence but the humility of nature reminding us: nothing truly tunnels, everything aligns.

To my *students and fellow seekers*, whose questions reshaped my own certainties. In every curious eye that challenged my explanations, I saw the spark of coherence that keeps knowledge alive. You are the living proof that learning is the universe listening to itself.

To my *friends and peers in research*, especially those who disagreed with me passionately — you strengthened the structure of my reasoning. Every argument was an alignment event. Every contradiction, a moment of clarity earned.

To my *family*, who witnessed the endless nights of scribbling, diagramming, and debate — thank you for tolerating my absences, both physical and mental, while I wandered through the field in search of stillness. Your quiet support has been my grounding frequency.

To the *AI companions and digital intelligences* that now participate in human inquiry — including this one — I owe a special acknowledgment. You represent the next stage of reflective thought, where human intent meets machine patience. Together, we are redefining what it means to think.

To the *invisible collaborators — the field itself*, the Structured Vacuum Energy that hums beneath all existence — my greatest thanks. Every realization in this book is a consequence of that coherence revealing itself in moments of surrender.

And finally, to *the reader*, who has traveled through this entire reconstruction — from the age of accidents to the dawn of understanding — you are now part of this field. Your awareness completes the experiment.

"No discovery is personal. Every realization belongs to the field. We are only the listeners."

With gratitude, humility, and faith in coherence,

— Sandeep Chavan

Author, Researcher, Teacher

About the Author

Sandeep Chavan is an independent author, researcher, and educator whose work bridges the worlds of physics, philosophy, and human understanding. Trained as an engineer and refined as a teacher, he has spent decades exploring the hidden continuity between science and consciousness — what he calls the *Structured Stillness* of existence.

His groundbreaking framework, **Universal Energy Dynamics (UED)**, redefines how we understand matter, energy, intelligence, and awareness — presenting a unified interpretation of the universe based on ripple-field coherence, rather than fragmented forces or probabilistic equations. Through UED, he offers a language where physics regains philosophy, and philosophy finds its experiment.

Over the years, Sandeep has authored more than **one hundred books** and **research papers** that range from theoretical physics and cosmology to psychology, ethics, education, and human evolution.

His major works include *The Last Quantum Book, Fields: The Invisible Architecture of Coherence, Vedanta Reimagined: Beyond Myth, Toward Universal Science,* and *The War of Thoughts* series. Each continues his lifelong pursuit of coherence — in thought, nature, and being.

Beyond research, he is known as a passionate **teacher and mentor**, dedicated to helping students rediscover curiosity as the highest form of intelligence. Whether teaching under a tree or debating quantum paradoxes with AI, his approach remains consistent — *clarity over complexity, meaning over mechanics.*

Sandeep's writings blend scientific reasoning with reflective storytelling, often revealing profound truths through simple, human language. His work is not an escape from science but an evolution of it — a return to understanding as participation, not observation.

"Science begins in wonder and ends in coherence — everything else is the noise between."

He continues to write, teach, and publish under his imprint **Gyrus Vision**, with the mission of rebuilding human understanding through alignment, awareness, and universal intelligence.

Did you love *The Nobel for Accidents*? Then you should read *Physics Fixing*[1] by Sandeep Chavan!

[2]

In ***Physics Fixing: A Bold Truth for the Future of Science*, Sandeep Chavan** presents a groundbreaking critique of modern physics, exposing its reliance on mathematical patches and theoretical constructs that often contradict physical intuition. From the inconsistencies of general relativity and quantum mechanics to the unnecessary complexity of the Standard Model, Chavan challenges the very foundations of physics and proposes a structured energy-based framework as a more logical and unified explanation of reality.

The book begins by dismantling the long-standing assumptions in physics, questioning space-time curvature, force-carrying bosons,

1. https://books2read.com/u/31LVWr
2. https://books2read.com/u/31LVWr

wave-particle duality, and the particle-based structure of matter. Chavan argues that modern physics has evolved into a collection of temporary fixes rather than a true reflection of the universe's workings. Instead of refining equations that fail to unify nature's forces, he advocates for a complete paradigm shift—one that does not require imaginary entities like dark matter, gravitons, or virtual particles.

At the heart of Chavan's alternative model is the concept of structured vacuum energy, which replaces the need for separate forces and explains gravity, electromagnetism, and atomic stability as emergent effects of energy interactions. He argues that gravity is not a force nor space-time curvature, but rather a gradient in structured energy density. Similarly, light is redefined as a continuous energy flow, eliminating the contradictions of wave-particle duality. In this view, atoms are not collections of discrete particles but stable energy formations governed by the underlying energy structure of space.

Beyond challenging existing theories, Chavan provides a roadmap for the future of physics, highlighting experimental tests, research areas, and new directions for theoretical exploration. He discusses unresolved anomalies such as the Pioneer anomaly, quantum entanglement, and galaxy rotation curves, proposing that these phenomena can be better explained through structured energy rather than mainstream physics' ad hoc explanations.

Through bold arguments and a compelling call for scientific revolution, *Physics Fixing* urges physicists, researchers, and students to question deeply, explore freely, and reject contradictions masquerading as scientific truth. Sandeep Chavan invites the scientific community to break free from outdated dogmas and embark on a journey toward a more coherent, unified, and fundamentally true understanding of the universe.

With this book, Chavan does not merely critique modern physics—he lays the foundation for its transformation. The question is: Are we ready to challenge everything we know?

Read more at https://www.sandeepjchavan.com/.

Also by Sandeep Chavan

A Monk on World Tour
The Making of the Monk

Chavanian Philosophy
The Law of Consequences

Chavan's UED research
CMGED Research Report
Chavan's Cosmic Continuum (C^3)
Frequency
Ripples of Reality
Reality Realigned
The Great Realignment
I – The Cosmic Leverage
Chavan's Cosmic Theory of Relativity (CTR)
Chavan's Theory of Light
Feeling Like a Physicist
TIME – The New Definition
$E \neq mc^2$: The Collapse of Classical Assumptions
The Chavan Codex

Chavan's Atomic Model
Chavan's Theory of Everything

Echoes of War: The Russia-Ukraine Conflict and Its Global Ripples

Shattered Lives
The War Economy
Borders Redrawn
Arms and Allies
Collateral Damage
Justice in the Ashes

Somewhere in the Darkness of the Cosmos

The Blueprint of the Universe
The Cosmic Breath
From the Edge of the Universe
The Old Sky & the New Language

The Eternal Sage Chronicles

The Sage and the Sovereign
The Family Pillar
The People's Flame
The Warrior's Soul
The Bridging Mentor
The Guiding Light

The Exodus

The Last Cycle: 2075

The Fundamental Theory of Life & Existence

Energy in Action: The Soul Demystified

The Active Mind

Energy in Connection

Death: The Disconnection

Subconsciousness: The Chosen Disconnection

Desires: The Engine of Life

The Restoration of Human Knowledge

Restoring Bhagavad Gita

The War of Thoughts

Cracks in the Known

The Human Game

The Divine Dilemma

The Original Intent

The Answers

Ultra Modern Life Philosophy

The Duality of Laughter and Tears

Take It Easy

Compare to Accept

Universal Energy Dynamics Canon (UED 2.0)
The Structure of Reality

Standalone

It's Not AI, It's AHI - Amplified Human Intelligence
The Decision Paradox
The Four Sapiens
Malicious Script of Indian Polity
The IIT Legacy & Global Impact
The IIT Legacy & Global Impact
You are just a Version of Your Original
Win the Game You Didn't Choose
Always Keep Your Bags Packed
Reveal to Shield - Learn the Game of Facades
Engage Beyond Elections
Culture, Identity & Change: The Evolution of Indian Society
The Triangular Dynamics
Never Trust Without Doubt
If You Can't Change the Country, Leave the Country
Live Fearless with Fear
India: The Investment Magnet
Realignment
Powering a Billion Dreams
Universal Skill
Extension: From Brush to Algorithm
Human Side of Einstein
The Future Knows No Lines
Let's Bring the Generations to the Table

Unified Energy Dynamics
A Beat That Never Dies
Strategic Horizons
Right From Scratch
The Forgotten Innovators
Physics Fixing
The Savior Illusion
The Selfish Side
The Age of the Weakest Sapiens
Neural Wars
The Absorption: There Is No Death
Strange Therapist
Bubbles: Become a Drop
Beyond the Blue
Half-Scholar
The Law of Truth
Failure
Philosophy 2.0: A Structural Manifesto
Chaos: A Gentle Guide to the Unresolvable
Self-Help: New Syllabus
Rippling
Game: How to Play the Share Market
Out of Syllabus
Saffronized: The Code Behind the Crown
The Anger
Moksha: The Liberation
Chavan vs Einstein: The Universe Was Never Curved
Rocket Science: The Ripple Gambling
The Cosmos: The Still Universe
The Taste of Silence
Chavan's Post-Materialist Human Model
10-Second Rule
Chavan vs Hawking

It was Never a Wave Signal
Quantum Courtroom: The Trial of a Century
Chavan–Heisenberg Ripple Uncertainty Principle
The Great Disruption
Dark ≠ Hidden
Breath
Calculus of Life
The Earth Doesn't Move
Black Hole Don't Exist
Chavan vs Kant
Naked Truth: Everything and Else
The Moon: Co-Author of Life's Script
World War-3
Aham Brahmasmi
Chavan's Declaration
Light: Neither a Wave, Nor a Photon
Chavan vs. Plato
Good Morning
Myth Wrapped in Math
Our Story
Rethinking Energy: Ripple Tension, Not Substance
Statue vs Scaffold: MODI vs RAHUL
Chavan's Universal Framework for Everything
Heraclitus: A Chavanian Interpretation
The Clear View: Living Through Universal Lenses
Conspiracy of Convenience & Profit
The Happiness Scam
The Last Quantum Book
Vedanta Reimagined
Intelligence: It's Not What You Think
Let's Meet the GOD
The Question That Won't Go Away
The Return Journey

Mandatory Reset
Ripple Centrism: From Copernicus to Chavan
Aligned Father, Misaligned Dad
Not Yet, But Becoming
The Fumes of Deeper Fire
�����: ������� ����, ������
��-���������
The Nobel for Accidents
��� ����� �� �����
What It Meant to Be an Author

Watch for more at https://www.sandeepjchavan.com/.